ENERGY AND INFRASTRUCTURE

Volume 3

The Fuelwood Trap
A study of the SADCC region

Full list of titles in the set
ENERGY AND INFRASTRUCTURE

The Fuelwood Trap
A study of the SADCC region

Barry Munslow with Yemi Katerere, Adriaan Ferf and Phil O'Keefe

publishing for a sustainable future

London • Sterling, VA

First published in 1988

This edition first published in 2009 by Earthscan

ISBN 978-1-84407-975-9 (Volume 3)
ISBN 978-1-84407-972-8 (Energy and Infrastructure set)
ISBN 978-1-84407-930-8 (Earthscan Library Collection)

For a full list of publications please contact:

Earthscan
Dunstan House
14a St Cross Street
London EC1N 8XA, UK
Tel: +44 (0)20 7841 1930
Fax: +44 (0)20 7242 1474
Email: earthinfo@earthscan.co.uk
Web: **www.earthscan.co.uk**

22883 Quicksilver Drive, Sterling, VA 20166-2012, USA

Earthscan publishes in association with the International Institute for Environment and Development

A catalogue record for this book is available from the British Library

Library of Congress Cataloging-in-Publication Data has been applied for

Publisher's note
The publisher has made every effort to ensure the quality of this reprint, but points out that some imperfections in the original copies may be apparent.

At Earthscan we strive to minimize our environmental impacts and carbon footprint through reducing waste, recycling and offsetting our CO_2 emissions, including those created through publication of this book. For more details of our environmental policy, see www.earthscan.co.uk.

This book was printed in the UK by CPI Antony Rowe. The paper used is FSC certified.

THE FUELWOOD TRAP

A Study of the SADCC Region

by Barry Munslow with Yemi Katerere,
Adriaan Ferf and Phil O'Keefe

*A study commissioned by the SADCC Energy Secretariat
and carried out by the ETC Foundation
(Consultants for Development)*

EARTHSCAN PUBLICATIONS LTD
LONDON IN ASSOCIATION WITH THE ETC FOUNDATION

First published in 1988 by
Earthscan Publications Limited
3 Endsleigh Street, London WC1H 0DD
Copyright © 1988 by The Technical and Administrative Unit of the SADCC,
energy sector (TAU) and the ETC Foundation

British Library Cataloguing in Publication Data
The fuelwood trap.
   1. Africa south of the Sahara. Fuel
   resources. Wood. Supply
   I. Munslow, Barry
   333.75
   ISBN 1-85383-007-0

Set in Great Britain by DP Photosetting, Aylesbury, Bucks

Earthscan Publications Limited is a wholly owned, but
editorially independent, subsidiary of the International
Institute for Environment and Development (IIED).

Contents

List of Tables, Figures and Boxes

Foreword

This volume emerged from work commissioned by the Technical and Administrative Unit (TAU) of the SADCC Energy Sector, to explore ways of tackling a growing energy problem within the region. Wood is the people's fuel and in a number of areas people are finding it more and more difficult to obtain the necessary supplies. This book describes a new approach to the problem, and its findings should help all those concerned with energy and development both within and outside of the SADCC region.

The main conclusion of the study, which was carried out by the ETC Foundation (Consultants for Development), is that the best way to ensure future woodfuel supplies, and simultaneously to prevent environmental degradation, is to improve the management of woody biomass within existing production systems based upon the innovations and responses already occurring among smallholder-farmers. For this operation to succeed at a national or even at a regional level, new relationships have to be developed between those ministries responsible for energy, forestry, rural development, environment and agriculture. These "new relationships" will be the basis for making the most of future wood supplies. Most importantly, these new relationships can be used to develop research and extension networks that will support woody biomass production by local farmers. This, of course, requires that strategies go beyond energy and forestry projects and are incorporated into as many other development schemes as possible.

Many people have contributed to the success of this work. In particular, a wide range of the region's own experts. Our thanks are due to all those who contributed their considerable expertise to the success of the project. The study was jointly financed by the Netherlands government and the European Economic Community, to whom thanks are also due.

J.T.C. Simões, Regional Co-ordinator, SADCC Energy Sector,
April 1988

Barry Munslow is Director of the Centre of African Studies, University of Liverpool.

Yemi Katerere is the Deputy Chief of Forestry in Zimbabwe.

Adriaan Ferf is the Manager of the ETC Foundation in the Netherlands.

Phil O'Keefe is a Reader in Geography and Environment at Newcastle Polytechnic.

Preface

This volume is the product of an enormous collective effort. It is based on a study commissioned by the SADCC Energy Sector and carried out by the ETC Foundation (Consultants for Development). The team which undertook the study was multi-disciplinary and multi-national but shared a commitment both to the general aims of the Southern African Development Co-ordination Conference and the specific task to which it was assigned – that of developing a fuelwood policy for the region. This book represents not only the work of the core team of professionals. Much more than this, its success (if such it is judged) has resulted from the enthusiastic commitment of the various other professionals within the SADCC region who played such a vital part. The most dedicated professionals to whom we owe our greatest debt are undoubtedly the smallholder-farmers who taught us the essentials and, most of all, made us aware that their knowledge and experience is the only starting point from which a real and indigenous development plan can begin.

Throughout the study we worked closely with Mr N.J. Salvador, Director of Wood Energy, and others within the SADCC Energy and Technical Administrative Unit (TAU) in Luanda. To all those people who contributed so much, we express our profound thanks. A Steering Committee comprised of TAU and representatives of the Royal Netherlands government and the European Economic Community, guided the study throughout. Initial findings were presented for review by experts within the SADCC region on a number of occasions. A formal review of the study was held in Harare in August 1987.

The main author of this volume was Barry Munslow. Management and co-ordination committee members were: A.J.E. Ferf, P. O'Keefe and B. Munslow. The team members were: G.W. Barnard, J. Fenhann, B.K. Kaale, Y.M. Katerere, P. Kerkhof,

G. Leach, M.K. McCall, M. McCall-Skutsch. F.C.O. Marleyn, A. Millington, S. Moyo, C. Pennarts, P.H. Phillips, K. Prasad, P. Raskin, T. Schouten, J.G. Soussan, J.R.G. Townsend, A. van Gelder and J. Venselaar.

ETC Foundation (Consultants for Development) Netherlands,
Kastanjelaan 5, PO Box 64,
3830 AB, Leusden, Holland.
Telex: 79380
Fax: 033-940791
Telephone: 033-943086

ETC (Consultants for Development) UK,
39 Norfolk Street,
North Shields NE30 1NQ.

PART ONE

The People's Fuel

1. *Setting the Trap*

There is a myth that we know all of the answers to the problems surfacing in the wood sector. The truth is that we know some of the answers but in some cases we do not even have the questions as yet. (Adolfo Mascarenhas)[1]

THE SADCC REGION

In 1980, nine Southern African countries pledged to work together to tackle their development problems by increasing regional co-operation. Those countries were Angola, Botswana, Lesotho, Malawi, Mozambique, Swaziland, Tanzania, Zambia and Zimbabwe. Together they agreed to form a regional organization known as the Southern African Development Coordination Conference (SADCC).[2] Their broad goals were to increase regional self-reliance and reduce dependency, particularly on South Africa. In subsequent years, much progress has been made but this has been in the face of the bitter opposition from the white minority government in Pretoria who have carried out a concerted policy of destabilization against neighbouring states.[3] This has affeced every area of SADCC's activities and not least that of the energy sector.[4] The world recession has had an equally severe effect on the economies of the region[5] but in spite of these difficulties, SADCC has continued its work.

Part of the reason for its success in the face of such problems is the careful way in which the structure for regional co-operation has been set-up. There is none of the over-arching, centralized bureaucracy which caused previous efforts at regional co-operation in Africa to founder. Such was the case, with the ill-fated East African Community, for example. Rather, the emphasis is upon a clear division of responsibility between the member states, each of whom takes on a particular portfolio. Where possible, this division of labour is based upon the particular strengths which each country brings to the group. Hence Zambia, a major copper producer, is responsible for mining. Mozambique, with its outlets to the sea, has taken on transport and communications and Angola, as the only oil producer in the region, looks after energy.

This book emerged as the result of a study commissioned by the SADCC Energy Technical and Administrative Unit (TAU) based in Luanda. TAU works to encourage and develop co-operation in regional energy resources. The regional perspective gained by such studies as this can precipitate a sharing of experiences which accelerates the learning process in development work. It can also lead to the more efficient use of aid and investment resources as it can avoid an undue replication of schemes.

The nine countries of the SADCC region share a common history of colonial occupation. Seven of the states were British colonies or high commission territories, whilst the other two (Angola and Mozambique) were colonized by Portugal. There was a considerable white-settler presence in the region and African farmers were denied access to a substantial area of land. A common pattern emerged of adult males leaving their plots to work as migrant labourers either in their own countries or frequently in neighbouring states. This meant that the women were burdened with additional agricultural tasks and brought up their families alone.

European settlement, increasing land clearance for agricultural production due to the rising population, and a growing demand for timber in the mining industry and in the towns began to affect people's access to wood. Certain cash crops that were introduced took a heavy toll on the trees. Tobacco, for example, requires one hectare of open woodland per hectare of crop produced. As African farmers were displaced from the better agricultural areas, they experienced both land shortages and, in certain places, a growing scarcity of wood. Large forests were taken over for commercial wood production or conservation, and local people were legally denied access to their own, formerly free, reserves. Wood consumption in the areas worst affected outstripped the annual yield and woodstocks soon became depleted. Land shortages led to a more intensive tree clearance as the number of mouths to feed increased in proportion to the farm land available. The traditional practice of leaving land fallow in order to allow biomass to regenerate and soil fertility to be restored, broke down under the pressure and the process of environmental degradation began.

Previously there was a more harmonious and organic co-existence between farming needs and the forests. Under the traditional forest – fallow system, shrubs and small trees were cut down and a fire was made around the larger trees in order to kill them. The burnt vegetation produced valuable nutrients for the soil, including potash. Crops could then be sown for several years until the soil needed rejuvenation and the land was then left fallow. Secondary forest development took place and farmers seeded certain trees for their future benefit. Over the fallow period, nutrients were replenished and nitrogen was fixed in the soil by leguminous plants or bacteria. The physical structure of the soil was maintained, thus enhancing its chemical composition and reducing erosion. The scene was then set for a repetition of the farming cycle.

WOOD – THE PEOPLE'S FUEL

More than 60 million people live in the nine member countries of the SADCC region. The vast majority rely upon biomass in the form of wood, charcoal and crop or animal residues for their basic household fuel. Indeed, biomass accounts for between 50 per cent (Zimbabwe) and 90 per cent (Tanzania) of national energy consumption. As Table 1.1 shows, woodfuel accounts for four-fifths of the total energy consumption of the SADCC region, principally because it is the major fuel for domestic use.

Table 1.1 SADCC Energy Flow (1985): Final Energy Consumption

	PJ(10^{15} Joules)	*Per Cent*
Woodfuels	1100	79·0
Oil products	143	10·3
Coal	87	6·2
Electricity	63	4·5

SOURCE: SADCC Energy Sector, Angola.

From the current demographic growth patterns and the slow transition to other household fuels, it appears that fuelwood consumption.will continue to grow in absolute terms. By the year 2000, there will be well over 100 million people living in the region and the vast majority will still rely upon woodfuel for domestic energy use. Yet some areas are already experiencing difficulties in obtaining fuelwood. With supplies diminishing and consumption growing, the region faces a major challenge. Perhaps the most significant challenge is that in a number of countries, where urban energy consumption is still largely dominated by wood, it will equal or surpass rural consumption within the next 20 years as urban areas continue to expand rapidly.

Unlike other energy sources, wood is not a solely marketed commodity. It does not carry a full production cost because it is often freely available. This makes it even more difficult for policy-makers and planners to devise solutions to its growing scarcity. The woodfuel problem was neglected for too long, precisely because it was assumed to be an easily available free good, there for the taking – to cook, boil water, give warmth in winter, brew beer, provide light in the evenings, and as a social focus for the household. It is now being recognized that throughout the region, though not in every area, there are problems in obtaining reliable supplies of wood. Furthermore, the situation is worsening.

Unlike the "other" energy crisis, that of oil, the woodfuel problem did not immediately ring warning bells for governments in the form of balance of payments deficits and fuel shortages affecting the vital arteries of transport and industry. Instead it was a problem experienced in dispersed, but growing locations throughout the region. But rather like drought which also starts slowly, the impact of the fuelwood problem can have far-reaching effects. It is, in short, another symptom of the collapse of sustainable "rural" development. Most importantly, the fuelwood problem is a burden carried, and carried quite literally, by women, whose voice so often remains unheard but whose labours keep the wheels of everyday life turning. Of course, commercial energy supplies are vital for Africa's development. Yet wood is the energy source for survival – it is the people's fuel.

THE HEAVY COSTS OF DEFORESTATION

There are many costs associated with the decline in woody biomass supply. With male migrant labour being such a common feature of the region's economy, women are frequently left in the rural areas to carry out all the tasks that were previously shared with men. A diminishing supply of fuelwood automatically increases the demands made on women's labour-time, with all the negative effects that this implies for their other work in agricultural production, child-rearing and housekeeping. This inevitably affects their health as well as that of their families. The most important response to a growing fuelwood scarcity has been more labour-time spent in its collection. But this of course has its limitations and carries with it a heavy cost, both to the individual and to the nation. As Irene Tinker has argued, the real energy crisis is all too often the lack of women's time.[6] In Tanzania and elsewhere in the region little girls help their mothers collect wood as soon as they can walk.[7] Women's labour-time is a constantly undervalued resource but it is the backbone of the rural economy. With men away from their homes working on mines, plantations, factories and in the transport sector, most of the peasant-farming work is effectively carried out by women.

A recent study by Allan Low, on agricultural development in southern Africa, has shown how farmers often take advantage of improved technologies such as hybrid seeds, fertilizers and tractors, not to produce a greater surplus for the market but to release labour-time.[8] This is either employed in migrant labour, which produces a higher return than crops (with the exception of Malawi), or in the case of women is used to collect fuelwood and water. Hence by alleviating the burden of fuelwood collection, agricultural production can increase.

The necessity of devoting more labour-time to fuel collection is only one of the increasing costs of energy consumption. Fuel scarcity can lead to a reduction in energy use, which may mean that less nutritional food is cooked. Cecelski has commented on the close statistical association between per capita food consumption and that of fuel.[9] More generally, health may be affected by the use of less heat in winter, and domestic hygiene is affected by a greater reluctance to use precious stores of fuel to boil water. As the availability of fuelwood diminishes, scarce cash resources have to be used to purchase alternative fuel and overall standards of living might fall. Deforestation carries a heavy social cost.

There are also the wider costs to consider, as a decline in woody biomass affects the huge array of uses of this versatile resource. For the smallholder farmer, trees provide many of their vital requirements. These include timber to build their houses, barns and fences; fodder and rubbing poles for their cattle; pharmaceuticals; agricultural implements; protection against wind and water erosion; maintenance of soil fertility; provision of fruit; and a habitat for game which is hunted and provides an important protein supply. This list could be greatly extended, with trees furnishing everything from cosmetics to shade from the African sun. As the woody biomass supply diminishes so does the availability of all the artefacts that come from trees and alternatives have to be found, usually through the cash economy. The decline in per capita income for the majority of people in the SADCC region results from the general development crisis, the world recession, and the devastating impact of South African-backed destabilization on the regional economies. In such a situation, finding the cash to purchase the alternative energy sources to fuelwood as a free good, is a major additional burden.

Perhaps the most serious long-term cost of diminishing woody biomass supply is that of environmental decline with its long-term impact on agricultural production systems. A recent SADCC report underlined the "increasing shortage of fuelwood resources for rural and urban use and the consequently fast depletion of woody vegetative cover essential for environmental protection, soil and water conservation, and improving agricultural productivity".[10]

The fuelwood problem is only one symptom of the rapidly changing socio-economic dynamics within the region, which have led to a deterioration in the quality of life in many homes. By trying to tackle the symptom alone, a result of defining the "fuelwood problem" too narrowly, those concerned in dealing with this issue can easily fall into the "fuelwood trap".

THE FUELWOOD TRAP

Only in the past few years has awareness dawned about the growing fuelwood problem. Until the oil crisis of the early 1970s, when the Organization of Petroleum Exporting Countries (OPEC) managed to negotiate a massive rise in oil prices, little attention was given to the energy question at central government level. Generally, there would be nothing more than an individual parastatal responsible for electricity supply. Petroleum was handled, and usually controlled, by the private sector. With the oil crisis, which had a devastating effect on the eight non-oil-producing countries of SADCC, the energy question received more serious attention and energy departments began to be created. One of their first tasks was to draw up energy balances showing supply and demand. They always stressed the supply figures, however, and only included commercial fuels such as oil, electricity, coal and gas. Even the international agencies drew up country energy balances only containing commercial fuels and always with the emphasis on the supply side of the analysis.

As fuelwood was not predominantly a commercially supplied product, the problem of its growing scarcity remained hidden. A breakthrough came when attention was focused at last upon this other potential energy crisis.[11] It took a number of years, however, before this issue was taken up by governments and donor agencies. Then, for the first time, national energy balances of supply and demand began to be constructed to include fuelwood.[12] As projected plans were made, the figures suggested a growing gap between fuelwood supply and demand. The initial response was to concentrate upon the two sides of the problem as it appeared when framed within this gap analysis. On the one hand supply needed to be increased, and on the other demand had to be curtailed if supply and demand were to balance out. As we will see, there were a number of problems associated with this approach.

The first of these concerned the institutional division of labour. Supply enhancement tended to be handed over to the forestry departments and responsibility for economizing on fuelwood use fell to the energy departments. Whilst this appears to be a rational approach, without the necessary co-operation between these two departments and their direct involvement in both sides of the problem, it inevitably creates difficulties in developing dynamic and coherent policy and planning initiatives. New modes of co-operation and, indeed, a rethinking of responsibilities, are in order.

Interventions from the colonial period onward have focused on increasing supply through conventional forestry. There have been two dominant models. The primary emphasis in forestry production has been on engineering wood supplies in marginal areas by creating commercial forestry plantations for industrial purposes. The second model has been the legislative designation of large areas of indigenous woodland as forest reserves, especially for watershed management. In both cases, the effect has been to deny the people access to land and forest resources, with the foresters assuming the role of policemen. When confronted more recently with the need to solve the urban fuelwood problem, peri-urban plantations have been advocated and in rural areas, either "mini-plantation" management or communal woodlots. The previous aim of foresters was to grow commercial timber for poles or pulp. Now, a similarly narrowly defined product, a "fuelwood tree", is advocated under the same plantation regime to meet the challenge of an inadequate woody biomass supply.

While peri-urban plantations are technically viable it is difficult for them to compete economically with fuelwood produced elsewhere. Not only have the production costs to be paid, but the opportunity cost of using the land for forestry rather than for other purposes, such as agriculture or building, must be considered. Getting fuelwood from natural woodland only involves the costs of extraction and transport. Effectively it is a "free good", hence it is difficult for the price of fuelwood produced in peri-urban plantations to compete with that of fuelwood from rural areas. A further difficulty is that people have a cultural preference for indigenous species for their fuelwood, whereas peri-urban plantations produce fast-growing exotic species, generally eucalyptus.

In developing communal woodlots, two approaches have been tried. The first entailed central direction and the completely external management of the community mini-plantations. The most advanced example of such a scheme within the SADCC region is the Lesotho Woodlot Project. By 1985, after twelve years of operation, 5000 hectares, divided between 275 locations, had been planted. Although the project has been successful in growing trees, in their study of it Crush and Namasasu concluded that: "The general disinterest displayed towards existing woodlots seems to be an expression of limited peasant participation in the establishment and running of the lots. More correctly, it reflects the realities of the hardships of rural existence. Woodlots do not provide immediate benefits to the populace, a vital factor in the daily struggle for life.'[13]

An alternative approach, in Tanzania, is also directed centrally, but the village takes responsibility for creating the woodlot. Many villages produced small woodlots following political persuasion, but these are of symbolic importance only as, in the majority of cases, they have never been harvested and have not expanded in area. However, it is noticeable that people are now growing more trees around their homes. At first the nurseries and woodlots experienced a night-time seepage of seedlings. The people who ran them soon became reconciled to the fact that people wanted trees around their homes and began to provide seedlings for this purpose.

It is important not to dismiss the role of communal woodlots entirely, in some areas they have played an important role as a demonstration. Where there is a strong sense of community and a history of community action they may even have some success. Yet results have often been disappointing. On occasion technical difficulties have played a part. More generally there has been concern about the way in which benefits are distributed, worries that land might be confiscated and conflicts over the use of land.

There is a role for peri-urban plantations and village woodlots but they represent only a narrow band on the spectrum of opportunity for supply interventions to tackle the fuelwood issue. Additionally, energy ministries have tried to reduce the demand for fuelwood by promoting conservation or, in the energy-richer SADCC countries, by encouraging the substitution of electricity, coal or oil. There is much of value to be learned in the experiences gained from these conventional supply-enhancing, demand-constraining and fuel-switching approaches, as we will go on to show. Alone, however, they are inadequate in tackling the fuelwood problem. This is essentially because the problem itself has not been properly understood.

The fuelwood trap, into which governments and donor agencies fall, is to assume that they have identified an obvious problem and consequently there has to be a simple solution. Unfortunately, this is not the case.

What then are the problems associated with conventional supply-enhancement strategies? Supply-and-demand balances and projections hide a complex pattern of specific areas of surplus and deficit. Fuelwood shortages occur in pockets or mosaics of varying levels of stress. As a site-specific problem, fuelwood planning consequently requires a decentralized approach for intervention to be effective. Unless both the specific site and people most affected are identified correctly, then any action taken will not have a significant impact.

There is also a fundamental flaw in the logic upon which supply-and-demand fuelwood energy analysis is based. Growing stocks and their annual increments are measured against estimated consumption and the "deficit" is thereby calculated. Little attention is paid to changing land-use despite evidence that it is not the demand for fuelwood which creates deforestation but land clearance for agricultural production. One estimate suggests that between 1950 and 1983, this was responsible for 70 per cent of the permanent destruction of forests in Africa.[14] These figures parallel global figures from the Food and Agriculture Organization (FAO) which suggest that, especially in Africa, land clearance will be the major cause of forest clearance even into the next century. In addition, the estimates of tree stocks frequently do not include farm trees and other sources of woody biomass which can be used in the hearth. This is not to argue that national analyses of the supply-and-demand situation are not required, but rather that better ways are needed to arrive at these. In particular, there must be a sense of scale, a way to separate site specific problems from the broad situational analysis.

Secondly, and of vital and more widespread significance, energy-planning conceptions of the problem have often been at odds with people's perceptions. Local people do not necessarily think of fuelwood as a problem, or at least a priority given the array of difficulties they face. There are a number of reasons for this. As a woman's task, it is a woman's problem and her voice has carried little weight. Land clearance for agricultural production or settlement has often produced a surplus of wood in the short term, but this simply defers the problem in time. Given that wood is perceived as a free good in rural areas, this makes fuelwood production appear economically unattractive to the individual smallholder.

Fuelwood is currently seen as a by-product from trees which provide fruit, fodder, fibre, construction timber and other materials important in the production systems within which people live and work. It is pointless, therefore, to grow fuelwood as an expensive commodity when it is popularly perceived as being free and a residue product. But international aid agencies have kept trying to grow "residues" and to charge full production costs for them.

The fuelwood problem is real and worsening. There are areas of fuelwood shortage and energy poverty, the labour burden on women and the quality of domestic life is being affected, and of major concern in the longer term is the rapid environmental deterioration caused by deforestation, which threatens food production and even life-support systems.

Naturally, governments have turned to forestry departments to enhance fuelwood supply. Here, they run into a problem. Tradition-ally such departments have been mainly concerned with industrial plantations and the protection and management of forest reserves. In other words, foresters were concerned with the forests. Fuelwood, however, is gathered mainly from trees and shrubs outside the forest, often on or around the farm. "Woody biomass" which includes shrubby vegetation, live fences and residues, is therefore the critical resource to consider – not the "forest".

In our study we draw an important distinction between the "forest" and "trees outside the forest". The departure point for this distinction is the available land classification which clearly denotes gazetted forests where the institutional responsibility lies with the state, including areas of watershed management, other protected areas, and commercial forestry plantations. These are what we refer to as "forests". "Trees outside the forest", on the other hand, are woody biomass found on and near to farms, and natural woodland, savanna and pastoral areas.

The first task is to get the foresters to look beyond the forests, at least for some of the time. They can then begin to address the problems that smallholder-farmers face, and encourage them to grow more trees themselves. For this to take place, new attitudes and training schemes will be required.

Inevitably there is often a colonial legacy of mistrust to overcome, with peasant cultivators historically viewing the state and large estates as encroaching upon *their* resources by alienating them from *their* land. Indeed, the policing role was forced upon foresters and this has inevitably created initial barriers on their part towards expanding their activities outside the forest, and the preserve of commerce and the state. On the other hand, it may lead to scepticism on the part of peasant farmers towards "belated" overtures being made by the forestry profession towards their concerns. However, the evidence from within the region suggests that these initial handicaps can be overcome and progress can be made.

A second and related difficulty associated with the traditional forestry department ethos is that it has concentrated on the direct production of trees and seedlings at inevitably high costs. Given the size of the fuelwood problem and the cost factors involved, forestry departments can never afford to "solve" the problem by their own production efforts. Thus there is an economic rationale for the shift that is taking place towards encouraging farmers and communities to grow more trees themselves. The evidence from the SADCC region shows encouraging results when this is tried in the right way.

Some have pointed to crop and animal residues as a possible way of increasing biomass energy supply to the household. The findings from our study suggest that the use of residues is more significant than is generally realized in the calculation of national energy balances. Throughout the region they are used for kindling, for short-time cooking and as an occasional supplement to the fire. They are used after the harvest, when the labour demands of the agricultural cycle are at their peak and the time available for the collection of fuelwood is reduced. There are several reasons why residues are relatively unimportant in the fuel bundle: wood is generally available within the region; residues are an inferior fuel requiring greater fire management; they create difficulties for simmering and are considered socially inferior; finally, residues have other important farming uses, notably as fodder and as manure. The evidence from within the region suggests that as the scarcity of fuelwood increases, people do not move rapidly down the energy ladder, rather they improve their management of wood as a fuel. However, in rural areas crop and animal residues are an energy safety-net rather than a preferred alternative to fuelwood, and offer limited scope for energy-planning intervention. Yet there does appear to be scope for improved residue use in modern industry.

When considering the other side of the equation, restricting fuelwood demand, there are currently considerable limitations on what can be achieved. Many efforts have been put into improved stove technologies and charcoal-making technologies. However these have only reached the experimental stage in the SADCC region. While they may have a role to play as one of a package of measures to deal with the urban energy problem, they are unlikely to have any meaningful impact on tree conservation in rural areas. There are three major reasons for this. First, fires are often made from free materials, usually three large stones or in some places, old buckets with holes punched in them. Secondly, stove improvement relies upon enclosing combustion and therefore interferes with the multi-purpose function of fires. Finally, as in fuelwood collection, the response of women to a shortage is to increase the labour-time spent on fire management. For these reasons, there is unlikely to be a rapid take-up of improved stove technologies in rural areas given scarce cash resources.

As we will go on to argue, the potential for restricting fuelwood demand is essentially a strategy for urban rather than rural areas. Improved stoves along with fuelswitch strategies can play an important role in tackling the urban energy issue and we will explore these options extensively. Fuelswitch and improved stoves become an option in urban areas precisely because fuel has become a valuable commodity, and free access to wood and residues as fuel is more restricted. A further consideration, but one outside current policy discussion, is changing dietary habits. In urban areas the increased consumption of processed food leads to a lower demand for energy.

One final option which has been proposed is the development of new and renewable sources of energy. However these remain in the domain of experimental technology and are unlikely to make a significant impact on energy provision in rural areas in the foreseeable future.

How then is the fuelwood trap to be avoided?

REFERENCES

1. A. Mascarenhas, "The relevance of the 'Miti' project to wood-based energy in Tanzania", in A.B. Temu, B.K. Kaale and J.A. Maghembe (eds.), *Wood Based Energy For Development* (Dar es Salaam: Ministry of Natural Resources and Tourism, 1984) p. 32.
2. There is a growing literature on this regional organization, see: A.J. Nsekela (ed.), *Southern Africa: Toward Economic Liberation* (London: Rex Collings, 1981); A. Kgarebe (ed.), *SADCC 2 - Maputo* (London: SADCC Liaison Committee, 1981); SADCC, *SADCC Maseru* (Gweru: SADCC, 1983); and the annual reports for the official documentation on SADCC. Useful commentaries on SADCC include S. Amin, D. Chitali and I. Mandaza (eds), *SADCC. Prospects for Disengagement and Development in Southern Africa* (London: ZED Books and The United Nations University, 1987); A. Tostenson, *Dependence and Collective Self-Reliance in Southern Africa* (Uppsala: Scandinavian Institute of African Studies, 1982); C. Stoneman, "SADCC raises its political profile", *Journal of Southern African Studies*, Vol. XIII, No. 1, 1986; J. Hanlon, *SADCC: Progress, Projects and Prospects* (London: Economist Intelligence Unit, 1985). A useful bibliography of SADCC is E. Schoeman, *The Southern African Development Coordination Conference. A Select and Annotated Bibliography* (Braamfontein: South Africa Institute of International Affairs, Bibliographical Series No. 14, 1986).
3. Studies on the regional confrontation include: J. Hanlon, *Beggar Your Neighbours* (London: James Currey, 1986); P. Johnson and D. Martin, *Destructive Engagement* (Harare: Zimbabwe Publishing House, 1986).
4. B. Munslow and P. O'Keefe, "Energy and the regional confrontation in Southern Africa", *Third World Quarterly*, Vol. VI, No. 1, 1984.
5. B. Munslow, P. Phillips, S. Kibble, P. O'Keefe, P. Goodinson and P. Jourdan, "The world recession and its impact on SADCC", in P. Lawrence (ed.), *World Recession and the Food Crisis in Africa* (London: Review of African Political Economy and James Currey, 1986).
6. I. Tinker, *The Real Rural Energy Crisis: Women's Time* (Washington DC: Equity Policy Center, 1984).
7. E. Cecelski, *The Rural Energy Crisis, Women's Work and Basic Needs: Perspectives and Approaches to Action* (Geneva: ILO Rural Employment Policy Research Programme, Technical Cooperation Report, 1985).
8. A. Low, *Agricultural Development in Southern Africa: Farm Household Economies and the Food Crisis* (Portsmouth, NH: Heinemann, 1986).
9. E. Cecelski, op. cit.
10. SADCC, *Food and Agricultural Report*, 1986, p. 9.

11. E. Eckholm, *The Other Energy Crisis: Firewood*, World Watch Paper No. 1, World Watch Institute, (Washington DC, 1975).
12. The first in-depth scientific study to attempt this was carried out in Kenya. See P. O'Keefe, P. Raskin and S. Bernow (eds.), *Energy and Development in Kenya: Opportunities and Constraints* (Uppsala: Scandinavian Institute of African Studies, 1984).
13. J.S. Crush and O. Namasasu, "Rural rehabilitation in the Basotho Labour Reserve", *Applied Geography*, 5, 1985.
14. J. Spears, *Deforestation, Fuelwood Consumption, and Forest Conservation in Africa: an Action Program for FY 86–88*, (Washington DC: World Bank (restricted), 1986).

2. Avoiding the Trap

Our departure point for avoiding the fuelwood trap is to begin looking more closely at the factors that determine supply and demand. If we can understand these more fully, they may provide a better indication of how to tackle the problem. As we will see, such an approach can lead to a fundamental recasting of the problem. The way to deal with the fuelwood problem may not be through fuelwood solutions at all but rather, paradoxically, through finding solutions to other problems. These solutions may have a greater chance of being adopted and thereby guarantee a dissemination sufficiently widespread to ensure a meaningful impact. This is because such solutions can meet more pressing and fundamental needs, yet still indirectly provide the woody biomass needed for the hearths of the region. This means trying to examine the entire chain from the fuelwood resource through to the consumer.

DEVELOPING A NEW APPROACH

Let us begin by examining what determines the extent of the woody biomass *supply* available for fuelwood consumption. First of all there is the nature of the woody-biomass resource itself which is defined by both environmental potential and the land-management system.

Environmental potential is principally determined by climate, topography and soil. Yet human land-use practices have, over the ages, profoundly effected the environment. Modern agriculture takes matters further by assuming that the environment can be mastered rather than husbanded. The history and changing nature of these land-use practices have severely affected the woody-biomass resource. Among a number of competing influences are the dominant forms of production, be they arable or pastoral, and the related land-tenure characteristics. External influences such as the development of a mining industry or urbanization will also have profound effects. Another important feature is the level of intensity of tree management within the land-use system. In the following chapter we will explore these ideas further, suggesting a relationship between the intensity and nature of tree management and the dominant forms of production.

Secondly, there are the competing demands from the local community for a wide variety of uses for trees other than as fuelwood. We have indicated just a few of these already. Trees provide the timber to build houses, granaries and fencing for livestock. Each form of construction will require different properties of the wood, hence particular species will be preferred. Trees provide fodder for cattle and fertilizer for soil. They provide food, medicines, dyes, natural fibres and shade. They are highly significant in the religious cosmology of Africa's peoples. Trees are life in its many aspects. Remove them and the protection for Africa's fragile soils disappears. In addition to these local needs, there is the external commercial demand, generally for the urban market. This is primarily for construction poles but also for a host of other products, including fuelwood.

The third and, in a period of rapid transition, probably the most significant factor is the available access to woody-biomass resources. Access can be restricted by both physical and social constraints. Physical access will be determined by topography and the preferred types of biomass available. Distance is obviously important in determining the ease of collection. Social access to woody-biomass resources is a complex question. Customary practices exist concerning the control of common land resources. Colonization brought with it land alienation and a level of privatization. Large areas of communally owned land became the preserve of both large- and small-scale private farmers, or was taken by the state. As social differentiation increased, access for the disadvantaged to the source of woody biomass often diminished. Table 2.1 provides a checklist of the factors affecting woody-biomass resources.

Obviously, there will be great variation from place to place and between different sectors of the community governing the effective supply of fuelwood. This variation must be the critical departure point for the formulation of plans and project identification. Fuelwood consumption in both rural and urban areas is determined by energy needs and the use of alternative fuels.

All these factors determine the effective supply of fuelwood. By

Table 2.1 Factors Affecting the Woody Biomass Resource

1. WOODY BIOMASS RESOURCE BY LAND-USE SYSTEM
   - environmental potential
     - climate
     - topography
     - soils

   - land-use practices
     - dominant production form
     - land tenure characteristic
     - external influences
     - intensity and nature of tree management

2. COMPETING DEMANDS
   - alternative uses by local community
     - construction
     - fodder
     - environmental protection
     - shade
     - food
     - cultural
     - medical
     - other

   - external commercial demand
     - urban woodfuel market
     - non-fuel demand
     - construction poles

3. ACCESS TO WOODY BIOMASS RESOURCES
   - physical access
     - distance
     - topography
     - biomass type
     - population density

   - social access
     - land distribution
     - customary practices concerning control of common land resources
     - enclosure of common lands for agriculture
     - state reserves
     - social differentiation
     - household decision-making (gender)
     - by social group
     - access via the market

"effective supply" we mean that proportion of the woody biomass resource which is available for fuel. Given these various and complex factors, woodfuel availability will inevitably differ enormously from place to place, reflecting the great variety in environmental potential and land-use systems, and the variety of competing demands upon the woody-biomass resource. Fuelwood availability will also differ between the different sectors of the community reflecting differing access.

Fuelwood *demand* is a reflection of energy needs, mainly for cooking and domestic heating. Population size, the technologies employed, the monetary or non-monetary costs of obtaining and consuming fuels, and the availability and costs of alternative fuels will determine this demand. Clearly there will be a considerable difference between rural and urban energy consumption in these respects.

The factors affecting woody biomass supply and those affecting fuelwood consumption (energy needs and the use of alternative fuels) are areas which must be identified when possible interventions are being determined. These factors can be drawn together into a system of fuelwood production and consumption. Figure 2.1 contains an outline of such a system and enables us to discuss viable points of intervention.

Looking first at the left-hand side of the diagram we see the crucial factors affecting fuelwood supply. The integrated production systems developed as a result of environmental potential and the dominant land-management system. This system determines existing land-use and the role that woody biomass plays both in and around the farm as well as in the natural forest and woodland. The best way to improve the woody biomass resource is by improving land-use practices based upon the particular set of tree and shrub requirements defined by the integrated production system. The nature of the integrated production system is defined by the particular configuration of productive activities employed to sustain rural communities. This is not a static phenomenon but will reflect human response to a bewildering variety of changes taking place. It is clearly possible to intervene to maintain woody-biomass supply or enhance the production of that supply in the integrated production system. To be effective, this will rely upon a proper understanding of the system and of the best ways of reallocating resources. The improved allocation of resources can be utilized to minimize the conflict between competing uses and to erode the access constraints.

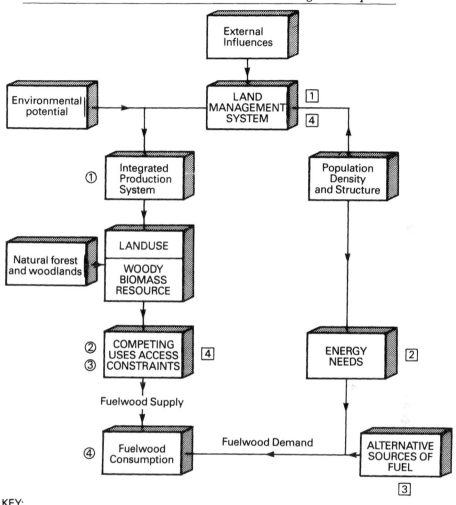

KEY:

○ Points of dislocation which lead to problems.
① Changes in the integrated production system which erode the biomass resource.
② Increased alternative demands upon the woody biomass resource.
③ Constraints upon access to potential fuelwood supplies.
④ Pressure upon fuelwood consumption.
☐ Points and main character of fuelwood interventions.
1 Supply maintenance or enhancement.
2 Conservation.
3 Fuel switching.
4 Resource allocation.

Figure 2.1 The System of Fuelwood Production and Consumption
and Possible Areas of Intervention

On the right-hand side of the diagram, we can isolate the factors that can reduce fuelwood demand. Conservation can be used to reduce overall energy needs and fuelswitching can make alternative sources of fuel available. The number of interventions likely to be effective, however, is relatively limited.

What becomes clear from examining the system of fuelwood production and consumption is that a different strategy will have to be developed for the rural and the urban areas. This is because the rural areas are a site of fuelwood production and consumption where neither the cash nor the distribution system are sufficient to permit widespread conservation or fuelswitching options on a scale large enough to reduce the fuelwood demand, given the rising population. Urban areas, on the other hand, are essentially a site of fuelwood consumption. Here, however, the cash is available, and a distribution and marketing network exists to enable fuelswitching, conservation and other initiatives to be taken. Given these different circumstances, we have divided this book into two distinct parts, dealing first with rural and then with urban areas.

The most significant area for intervention would appear to be improving the management of woody biomass within production systems. Realizing the limited scope for the direct production of fuelwood allows us to construct a new way of seeing the fuelwood problem. In any one area, the land-use patterns will provide a generalized picture of the opportunities to provide fuelwood. To intervene effectively in that given area, it is important to be able to interpret the existing landscape and to identify how this mosaic changes over time. It is the accurate reading of this changing mosaic that is required for successful intervention.

The starting point for this broader and more dynamic approach to the fuelwood problem is understanding the role of woody biomass within land-use systems. Trees and shrubs play an important part in the overall environment in which smallholder-farming families gain their livelihood. Of central importance to the farmers and their families is the agricultural production system. Woody biomass plays an important role in this system and the best way to encourage increased woody-biomass production is to demonstrate how this can enhance the overall productivity of the agricultural (and life-supporting) system. Put at its most stark, there is no point in devising a strategy to provide fuelwood for cooking if the food-production system has collapsed. What this means is that the place to start tackling the fuelwood problem is not with ready-made, narrowly defined "fuelwood" solutions. Rather we must listen to the smallholder-farmers' own priority needs within his and her production system and see how increased woody-biomass production can meet those needs.

As a by-product, the fuelwood will become available from dead branches and residues. In whatever form the wood might be used in the interim, as fencing, tools, etc., it can always be burned eventually. Trees and shrubs always produce wood residues for burning, whatever their primary function for a given household or community. This means that initiatives can be broader in their scope than growing "fuelwood" trees and this will help ensure maximum local support.

It is important to realize then that fuelwood is still largely regarded as a by-product of the woody-biomass management system. The creation of living hedges to demarcate land, the establishment of shade trees around the house and the deliberate production of a number of shrubs, bushes and trees for fruit, fodder, herbs and other yields is likely to provide the critical departure point for the encouragement of wood production. In time, there will undoubtedly be a growing commercial market for fuelwood *per se*, but until such opportunities arise, it is important that wood production be directed towards end-uses other than fuel. For example, the commercial opportunities presented by the demand for construction poles may be one logical departure point for wood initiatives in certain locations.

Thus the role of woody biomass needs to be identified and specified in its relationship to the wider production system. In developing this approach to the rural areas it is then possible to identify the problems with greater precision and the constraints and opportunities that exist. Such a focus also brings out the indigenous responses that have remained hidden from view.

LEARNING FROM THE PEOPLE

Throughout the SADCC region there are trees to be found: around the home, dotted in the fields and grazing areas, and in the hedgerows. Most smallholder-farmers plant some trees, although the extent of this practice varies from place to place. Hence there is already a tradition of woody-biomass management on which to build. This is not by any means a static tradition as farmers have learnt to adapt to changing circumstances. It is abundantly clear that there have been important indigenous responses to the management of woody-biomass resources within integrated land-use systems. Generally, people have developed their own principles of environmental-resource management based on their own experience of their own conditions, as Richards has forcefully demonstrated.[1]

Unlocking the storehouse of rural people's knowledge and finding ways of integrating this with modern scientific ideas is the key to tackling the fuelwood problem. But what exactly is this knowledge? To begin with, as Chambers has argued, there are farming priorities based upon a longstanding observation of the local environment.[2] Mixed cropping is an obvious example and the colonial emphasis on monocropping is now acknowledged to have been a mistake for small farmers in many areas. Knowledge of the local environment is unsurpassed and any outside plan or modern scientific technique introduced will inevitably have to face the rigours and idiosyncrasies of that environment. In the SADCC region, variations in altitude produce a wide range of ecological niches. A third important factor is the strength of rural people's ability to maintain, extend and correct local knowledge. Finally there is the experimental nature of local practices which can be both innovative and risk-minimizing. It should never be forgotten that the agricultural revolution in Europe was based upon science codifying rural people's knowledge and practices.

Whilst recognizing the importance of indigenous technical knowledge and people's active responses to their changing environment, it is necessary to emphasize that they cannot always cope. Outside support of the right kind can provide much needed help. The successful implementation of indigenous technical knowledge requires the right social context, in particular a stable community.[3] Powerful external forces of change can undermine this and diminish the spread of essential knowledge and skills.

None the less, the rationale for supporting indigenous technical knowledge and experimentation remains powerful. Leach and Mearns have ably summarized the threefold advantages.[4] First, it avoids segmenting the integrated reality of people's own experience of life into administratively defined executive compartments such as forestry, health care, agriculture and nutrition. This way, the causes rather than the symptoms of their problems can be better tackled. Secondly, given the unique specificity of conditions at the local level, outside researchers could not cover all the required terrain.

Hence farmer participatory research, in which the farmer is researcher not respondent, is both more cost-effective and more suitably adapted. Thirdly, this approach develops local capability, self-confidence and control, important developmental goals in their own right.

A key area of concern here is the ability of foresters to incorporate knowledge developed outside the sphere of organized "scientific" forestry. This demands an approach which values and includes traditional knowledge and practices in woody-biomass management. Robert Chambers has spelled out convincingly and in great detail the arguments for more generally adopting such an approach. He points out how,

> The links of modern scientific knowledge with wealth, power and prestige conditions outsiders to despise and ignore rural people's own knowledge. Priorities in crop, livestock and forestry research reflect biases against what matters to rural people.

But his message is one of hope not gloom, for he continues,

> Rural people's knowledge and modern scientific knowledge are complementary in their strengths and weaknesses. Combined they may achieve what neither would alone. For such combinations, outsider professionals have to step down off their pedestals, and sit down, listen and learn.[5]

By combining the two spheres of knowledge, first in their training and later in their extension work, foresters will acquire a broader range of skills and a greater capacity for being effective in a sphere of activities expanded to include trees outside the forest. Acquiring such skills is vitally important if they are to serve their new constituency. The historical legacy of foresters, as we have seen, is one of preserving the state's interests in protecting forest reserves and the interests of companies in large-scale commercial forestry plantations. In essence, the fuelwood challenge demands that foresters (along with energy planners, agricultural extension workers and many others beside) look after the interests of peasant farming households by facilitating the improved management of the woody-biomass component of their integrated production systems so that local livelihoods are sustained and enhanced. International recognition of the importance of local tree-growing has been increasing. As Flores Rodas, the Assistant Director-General of the FAO Forestry Department, has commented: "Programmes to encourage and support rural people in these efforts have become one of the principal tasks of forest services." [6]

It is slowly being accepted that successful development interventions must value the worth of indigenous responses, that is address peoples' own perception of their needs and ensure *their* participation, and that the developmental process must be one which is sustainable and therefore environmentally sound. In order to address this, it is necessary to adopt an integrated production systems approach. This is the perspective adopted by rural cultivators themselves and it therefore provides greater opportunities for intervening in ways that will be more readily acceptable and potentially beneficial to rural peoples.

Before the colonial land constraints developed, shifting cultivation and pastoral patterns of movement allowed fuelwood energy to be solely a *collected* item. As land constraints ensued, greater responsibility for the indigenous *management* of woody biomass followed. Hence farmers left certain trees for shade and other purposes when clearing ground for arable production. As the demands of population and cattle on a limited resource base grew, and the destruction of woody biomass increased as a result, the need for *production* initiatives in woody-biomass management became clearer. In the future, these will take on a growing importance. In certain ways this parallels the neolithic agricultural revolution which marked the change from food collection to food production. An understanding of these processes is vital for the formation of policy interventions, planning and project identification all of which must aim to develop a "neolithic" revolution for woody biomass, a move from the simple collection to the sustained production of wood.

To illustrate the problem more concretely, let us look briefly at the crisis experienced by rural peoples in many areas of the region. Increasing population pressure on scarce land resources (a result of colonial land divisions and alienation) leads to increased pressure on available land, inadequate grazing, overstocking, and the erosion of woody-biomass resources as well as the soil. As rural peoples depend upon arable and pastoral production for their subsistence (as well as for socio-cultural reasons) it is the woody-biomass component of the system that is frequently placed under threat causing traditional systems for the management of woody biomass to enter into crisis.

Agricultural production systems do not remain static but respond to changing circumstances. Tree-management practices also change in response to new influences and pressures. Some of the changes that are occurring increase the incentive for growing trees. Privatization of land and the penetration of commercial markets for firewood and poles both provide new opportunities for planting trees that were not previously there. Other changes make tree-growing harder and effective management of natural woodlands much more difficult.

However, the point is that rural people are rarely helpless. They have had to deal with changing circumstances before, and they will have to do so again. Their resourcefulness and adaptability are often far greater than is generally assumed.

Ester Boserup has produced an important explanation of why changes in agricultural production systems occur.[7] As a population rises and land availability decreases, there is a tendency to move from extensive to intensive production systems. When land is freely available, the returns for labour are comparatively higher under extensive cultivation practices. It is a growing land shortage, therefore, that compels the adoption of innovative intensive cultivation practices. The lesson here is that *people under pressure respond positively to change.* As we will go on to show in chapter three, the most developed agroforestry practices are found in the areas of high potential where natural woodland has been severely reduced and there is a dense population and intensive agricultural production.

Richards, however, has produced some important qualifications to the Boserup theory, stressing that it is a "model" rather than history based upon detailed evidence, and may be misleading as a basis for formulating agricultural development policies.[8] In other words, if techniques of shifting cultivation are simply seen as an early stage in the evolution of agricultural production systems then, logically, development should be concerned with getting rid of them. Richards points out that if labour-intensive innovations come about through population pressure, and shifting cultivation was the original preference, this does not necessarily signify that one is superior to the other. Indeed, Geertz and others have argued that these changes are "involutionary", meaning that it is a case of running faster just to stay still.[9]

Richards prefers to see shifting cultivation as the rich tool-kit of land-management practices and he rightfully argues that this approach "allows for the possibility that individual skills and specific elements derived from shifting cultivation may still be of importance even if the overall farming strategy needs to be changed."[10] In other words there is no point in throwing the baby out with the bathwater.

What follows from the approach we have outlined, is that the starting point for tackling the fuelwood problem must be the multiple roles that trees perform in the environment within which people live and work. Starting by establishing people's needs, energy planners can then go on to look at what trees would be suitable for a particular area and how long they take to grow. These decisions enable the woody biomass production system to be developed and managed. By understanding and incorporating the knowledge of the people, they can make available a rich toolkit for the management of woody biomass and can explore new avenues for a research and development programme led by the needs of the people.

The time factor is important. While government-run forestry plantation strategies have a relatively long timescale, the same is not true for the small farm. The cycle of production has to be short enough to make the options attractive to the small farmer in addition to addressing the particular tree end-uses required. Moreover, this timescale must fit in with the risk-minimizing strategy of peasant farmers who have no guaranteed price-support mechanism to sustain their income. Circumstances have to be created which encourage the planting and growing of species to meet these needs and these species have to be made available.

It is frequently in dry areas of communal tenure where the largest concentration of the poor are found and where the pressures on the woody-biomass system are the greatest given a shortage of land. This frequently leads to the unauthorized but often harmless cropping of woody biomass from the adjacent natural woodland in neighbouring commercial areas (the same may hold true for grazing) and forest reserves. As we will go on to show, there is plenty of scope to expand upon resource-sharing options in a more organized manner.

CONCLUSION

We have suggested a new way of looking at the fuelwood problem, focusing on the place of woody-biomass management within the agricultural production system. This new approach is of most obvious benefit to the rural areas and we will explore its implications in greater depth, in the following section of this book. In the urban areas, the problem is rather more complex and here a number of different approaches will be explored.

We strongly argue that separate policies are required for rural and urban areas in order to meet the fuelwood challenge effectively. The reason for this is essentially that rural areas are a site of fuelwood production and consumption with few viable alternative energy supplies for domestic use. Urban areas, on the other hand, are a site of fuelwood consumption alone where alternative fuels and the cash to purchase them are more readily available. In Part Two, we examine the rural areas and what people are already doing to manage the woody-biomass resource. We emphasize in particular, the importance of woody biomass in guaranteeing a future for development schemes which can be sustained.

REFERENCES

1. P. Richards, *Indigenous Agricultural Revolution* (London: Hutchinson, 1985).
2. R. Chambers, *Rural Development. Putting the Last First* (London: Longman, 1983). See Chapter 4 for an elaboration of these arguments.
3. J. Farrington and A. Martin, *Farmer Participatory Research: A Review of Concepts and Practices*, Discussion Paper 19 (London: Overseas Development Institute, 1987).
4. G. Leach and R. Mearns, *Bioenergy Issues and Options for Africa*, A Report to the Royal Norwegian Ministry of Development Cooperation (London: IIED, draft 1988).
5. R. Chambers, op. cit., p. 75.
6. Foreword to FAO, *Tree Growing by Rural People*, FAO Forestry Paper 64 (Rome: FAO, 1985).
7. E. Boserup, *The Conditions of Agricultural Growth: The Economics of Agrarian Changes under Population Pressures* (London: Allen & Unwin, 1965).
8. P. Richards, op. cit., pp. 51 onwards.
9. C. Geertz, *Agricultural Involution: The Processes of Ecological Change in Indonesia* (Berkeley: University of California, 1963).
10. P. Richards, op. cit., p. 55.

PART TWO

Rural Areas

3. Building a Sustainable Future

AFRICA'S CRISES

A recent study by Paul Harrison has highlighted four essential aspects of the crisis facing Africa: a decline in food production per capita, increasing poverty, rising debt and finally, and most serious for the future as it will profoundly effect the other three, the environmental crisis.[1] We will argue that trees and woody biomass more generally, are too important to be consigned "merely" to a fuelwood problem. Developing a policy for woody-biomass management can positively affect all four aspects of Africa's crisis because woody biomass protects Africa's soils and hence its productive potential. It also provides many of the needs of Africa's peoples. Annually, 3·7 million hectares of woodland and forest disappear, and already more than a quarter of the continent is undergoing desertification ranging from moderate to severe.

The seriousness of the problem is enormously magnified by the fragility of Africa's soils. Spanning the equator, like no other continent, Africa receives the assault of the sun and the battering of a heavy but erratic rainfall. Given the poverty of the people, this climate produces disease and pests in abundance. In many areas sleeping sickness (trypanosomiasis), for example, makes the use of draught animals for the intensification of agriculture difficult. Biomass cover gives protection from the sun and rain to the continent's fragile soils. Remove it and the climate wreaks a terrible revenge, soils are washed away and fertility is lost. So, maintaining biomass cover in general and woody biomass in particular, is absolutely vital to a sustainable African environment. Therefore woody-biomass management is essential for the long-term sustainability of peoples' livelihoods in Africa.

Yet why should people plant trees? What concrete benefits are in it for them? The majority of Africa's peoples are the poor who live in rural areas. They face a daily struggle for survival. Meeting their most basic needs on a day-to-day basis is all that can concern them. How can providing for a sustainable future be encouraged when simple survival is the order of the day?

SUSTAINABLE DEVELOPMENT

The answers to the above questions would seem to lie in finding those solutions to people's immediate or short-term needs which will also have the longterm benefit of maintaining the ultimate source of their livelihood – the environment. The World Commission on Environment and Development, commonly known as the Brundtland Commission, arrived at a useful definition of sustainable development which summarizes this approach:

> Sustainable development is development that meets the needs of the present without compromising the ability of future generations to meet their own needs.[2]

The issue of woody-biomass management, then, is much more important than the fuelwood problem alone. It has a vital role to play in guaranteeing the sustainability of current and future life-support systems.

Land degradation carries a high cost to the development effort. Either more labour and capital are needed to produce the same return, or a decline in production occurs if both these inputs remain constant. The concept of land degradation must be socially defined, in that it denotes a declining social use gained from the land. In other words, land degradation means a loss of capability to satisfy the demands made upon it.

Land-use need not always have a harmful effect, however, as it can increase productive potential. There are both positive and negative processes at work. The net degradation that results from a given land-use can be represented by the following relationship, as Blaikie and Brookfield have demonstrated.[3]

Net degradation *equals* (natural degrading processes *plus* human interference)

minus

(natural reproduction *plus* restorative management)

This signifies that on the negative side, degradation is created both by natural processes of erosion and by the acceleration of those processes by human interference which transforms land-use. On the positive side, traditional land-use practices were based on ensuring an ecological balance and sustainability by creating a forest – fallow system which enabled a natural reproduction to occur, restoring soil fertility. As the pressure on the land resource increased under colonial and post-colonial circumstances, the imperative for restorative management also grew. In other words, it was no longer natural enough to use regenerative capacities alone. A more active and interventionist practice was required, based upon both the resource of local knowledge and the admittedly belated application of Western scientific research.

Land-use management, therefore, should be concerned with

> applying known or discovered skills to land-use in such a way as to minimize or repair degradation, and ensure(s) that the capability of the land is continued beyond the present crop or other activity, so as to be available for the rest.[4]

Figure 3.1 The Relative Sensitivity and Resilience of Land and its Implications for Land-use Management

	Low Resilience	*High Resilience*
Low Sensitivity	Initially resistant to degradation but once a certain threshold is reached, difficult to restore capability	Fairly immune to degradation. Land--use management interventions can have rapid positive effect
High Sensitivity	Easily degrades and does not respond to improved land-management	Suffers degradation easily but responds well to improved land-use management

SOURCE: Derived from Blaikie and Brookfield, 1987.

Two important concepts here are the *sensitivity* of a land system following human activity and the extent of the changes set in motion by the interaction of natural forces; and *resilience*, that is the ability of land to regain its capability after interference. Figure 3.1 indicates the array of situations that it is possible to find given the low or high *sensitivity* and *resilience* factors of land and its subsequent implications for land-use management. This matrix can aid the identification of constraints and opportunities for intervention. Improved woody-biomass management, however, is only one of an array of improved land-use management practices available. But we would argue that it is one which shows great promise and, furthermore, it is one which is widely known and practised already.

Richard Franke, in his work on the Sahel, has made two interesting observations.[5] The first is that in the frontier zones where herders and farmers are in intensive contact, exchanging knowledge of their animals and plants, the most ecologically sound Sahelian production systems are to be found. The basic argument here is that mixing, and experimentation provide a broader range of alternative systems and therefore greater opportunities to cope with the hazards that are encountered.

Franke's second observation is that the most ecologically sound Sahelian production systems arise where the producing classes have had substantial power vis-à-vis the dominating classes. Hence where ruling classes arose in pre-colonial Sahel, their increased demand for output was often in conflict with resource-maintaining practices.

> The spread of modern power and class relations was thus most likely correlated with the decline of indigenous resource protection practices, except in those cases where producers escaped to the fringes of the empires and could maintain control over their own labour and reap its benefits themselves, or ... where an organized movement for redistribution and a lessening of the demands of the rulers could make use of a similar strategy.[6]

What the Sahelian experience implies for Franke is that the best results are likely to emerge where the management and control of land, labour and technology are by the people themselves – those who have the greatest interest in protecting and developing their own resources. Yet the ownership or rights to those resources have always been a subject of dispute in the transitional process.[7]

It is clear that indigenous knowledge alone is not sufficient to deal with the problem, otherwise there would be no need to write this book. More knowledge and research is required and above all the necessary resources must be made available to tackle the land degradation problem adequately. Beinhart, in his study of the historical experiences in southern Africa, concluded:

> although there is considerable evidence of self-regulatory practices in landuse in pre-colonial African societies, it cannot necessarily be assumed that peasant communites in the colonial period, their old systems of authority eroded, and faced with a shortage of land, increasing population densities, new opportunities and new constraints in their battle for survival, had the capacity to regulate themselves.[8]

Colonial rule was to massively disrupt Africa's social and economic structures. Land-use-management practices were inevitably affected. The most detailed record of this impact is the study by Helge Kjekshus of Tanzania. This shows how the sophisticated and well-established system of ecology control was severely broken down. Until the end of the nineteenth century, Tanganyika's people lived, in the words of the explorer Burton, in "comfort and plenty". The onset of colonial rule in the final decade of the century brought with it a series of calamitous changes resulting in disruption in ecological control. Sleeping-sickness epidemics, for example, resulted from a rapid depopulation of people and cattle with an attendant loss of control over the environment, pushing many people "back to a frontier situation where the conquest of the ecosystem had to recommence".[9] Colonial land annexation obliged people to settle in areas they had previously avoided and this encouraged the spread of tsetse fly, smallpox and rinderpest diseases.[10]

Studies in other parts of the region confirm these disruptive effects. For example, Terence Ranger has outlined how in Zambia the reserves of the Swaka were so reduced by 1929 that the traditional fallowing periods of the *citamene* system could no longer be employed and land degradation ensued.[11] In eastern, central and southern Africa, European annexation of land and the creation of migrant-labour economies had a serious negative effect on land-use management. With men usually away for periods longer than the annual agricultural cycle,[12] there was less labour available on the farm and so conservation practices were more difficult to pursue. Furthermore, as many of the traditional practices presupposed a lower population density and greater land availability, these practices broke down in areas of high land-use stress.

Yet increasing population pressure, as Boserup has argued, can provide the stimulation for innovatory practices.[13] Land-use management practices are not a static phenomenon in African societies and this century has seen positive responses from among the farming community to the rapid change that is taking place.

From this broad discussion of the general problems of land degradation and sustainable development, we will turn to focus on people's existing knowledge of tree growing and management. But first we will outline the broad spectrum of land-use management strategies that can be found within the SADCC region, as this appears to influence the levels of intensity of tree-management to be found.

LAND-USE SYSTEMS

Four strategies of land-use are readily apparent in the landscape. The first is largely based on sedentary, rain-fed agriculture. This strategy, which we define as *arable* farming is characterized by:

1. The intensive use of small sites.
2. The relative immobility of people.
3. Short seasons of concentrated agricultural work largely dependent on rain, which produces a seasonal "unemployment" contributing to a search for non-agricultural jobs in urban areas.
4. Cultivation of specific plants for defined end-products.
5. Supplementary multiple use of natural vegetation.
6. Minimal livestock kept for household products.
7. A focus on market production.

The second land-use strategy is based on *pastoralism*. The characteristics of this system are:

1. The extensive use of large areas for grazing.
2. High mobility of people and livestock.
3. Year-long land-use practices.
4. Multiple exploitation of plants, but little cultivation.
5. Little focus on market production.
6. Animals as the focus for wealth accumulation.

Across SADCC member-states, there is much variation in local strategies of land-use, frequently combining elements of both the arable and pastoral systems in a third *mixed farming* strategy. In addition there is a fourth land-use strategy, usually controlled by the state, which seeks to limit the human impact on natural woodland and vegetation by restricting agricultural land-use opportunity. Watershed conservation and game-parks are two well-known examples of such strategies.

Clearly, each of these land-use systems will give different priorities to, and make specific demands upon, the woody-biomass resource. We also find that the intensity of tree-management practices will differ and this can be related to the various systems of land-use. All too frequently, however, people's knowledge of tree-growing has been ignored as a potential development resource.

PEOPLE KNOW HOW TO GROW TREES

Managing the woody-biomass resource is not a new phenomenon in Africa. The traditional forest–fallow system previously described, was a complex management system based upon careful observation of the environment. An early essayist who provides us with the historical evidence of active tree-management within the region at the turn of the century, was the German, Brusse. He found tree-planting to be an integral part of the agricultural system in two areas, Matengo and Ungani, in the south-west of Tanzania.[14] In East Africa, mangoes were initially planted along slave routes to feed captives on their way to the coast. We learn from the historical record that in the precarious climatic conditions of northern Namibia, Ovambo kings, "strictly preserved the fruit trees and checked excessive deforestation".[15]

Although the written historical record on this subject is very poor, more recent surveys demonstrate that smallholder-farmers are indeed actively managing the woody-biomass resource by planting trees. A survey undertaken in Zimbabwe in 1984, found that 70 per cent of the households questioned, indicated that they planted trees.[16] Two-thirds of the trees grown were to provide fruit and almost a third were of eucalyptus, primarily for the provision of building poles. Interestingly, well over half the seedlings were either self-grown or collected wild, indicating a very active woody-biomass management. Sixty-six per cent of the households planting trees did so on their own initiative. The evidence from this survey confirms that there is a hitherto untapped potential of woody-biomass management existing within the rural community.

A recent FAO Forestry study concluded that:

Almost everywhere, a certain standing stock of different types of trees, whether deliberately cultivated or allowed to grow naturally, has been recognized as necessary by farming communities. ... The many products and benefits which rural people derive from trees reflect detailed and sophisticated knowledge about their immediate environment. The assumption that traditional communities are unaware of the benefits provided by trees, and therefore need to be educated about the immediate consequence of the depletion of tree cover, is rarely accurate.[17]

Until recently, the woody-biomass component in farming systems was almost entirely overlooked. Little was understood about the complex role that trees play within production systems. Information on the types of trees grown remains scarce. Even less is known about why farmers prefer certain species, what methods they use in growing them, and what factors lie behind their traditional practices.

This kind of information, however, provides vital clues as to how people might be encouraged to grow more trees by setting up projects that build upon existing initiatives or by integrating a woody-biomass component into development projects not primarily concerned with trees. It is important for all those involved in these issues to appreciate the nature and extent of indigenous woody-biomass management practices.

TREE MANAGEMENT SYSTEMS

For the sake of clarity, a distinction can be drawn between three broad categories of tree-management systems:

- high-intensity tree-management systems
- medium-intensity tree-management systems
- low-intensity tree-management systems

As we will go on to show, these demonstrate a marked correlation with the arable, mixed-farming and pastoral production systems.

High-Intensity Tree-Management Systems

In parts of the SADCC region, traditions of tree-management are highly developed. Although these cases are the exception rather than the rule, there are several areas where large numbers of trees are

deliberately grown by local farmers for a whole variety of end-uses.
The most dramatic examples are found in high potential arable-farming zones where soils are good, rainfall is plentiful and population densities are high. With the majority of the indigenous forest long since cleared, local populations have had to develop their own alternative sources of firewood, timber and other tree products. In the traditional "chagga" farming system, practised on the southern and eastern slopes of Mount Kilimanjaro in Tanzania, trees form a major part of the farming system. They are grown together with food crops and shrubs in a multi-storey, multi-product "home-garden".

Dividing them horizontally, the home-gardens can be separated into a number of levels:

- the lowest level (0–1 metre) consists of food crops such as taro and beans, along with various herbs and grasses.
- the next zone (1–2·5 metres) consists mainly of coffee bushes with a few young trees and shrubs, and medicinal plants.
- next is the banana canopy (2·5–5 metres) with some fruit and fodder trees.
- above this, the zones are less distinct, with a diffuse zone (5–20 metres) of fuel and fodder species.
- the uppermost zone (15–30 metres) is of valuable timber trees, together with some fodder and fuelwood species.

The types of tree grown here include both indigenous and exotic species. Some of the most common are:

Albizia schimperiana	–	fuelwood, building material
Cassia didmyobotyra	–	medicinal uses
Cordia abyssinica	–	shade for coffee, fuelwood, building material, beehives, construction
Croton macrostachys	–	fuelwood, fodder, insect-repellance
Grevillea robusta	–	fuelwood, building material, shade for coffee
Markhamia platycalyx	–	termite-proof building poles, fuelwood
Morus alba	–	fodder, fuelwood, fencing
Tectona grandis	–	high-quality timber

Similar traditions of multi-storey tree cultivation are also found in Bukoba District, to the west of Lake Victoria. Here *Grevillea robusta* is widely grown to provide shade for coffee, with timber and firewood being important by-products. This practice was introduced in early colonial days. Since then, it has gone through several cycles of popularity and disfavour – at one time in Tanzania, the

British administration tried to ban it in an attempt to increase coffee yields. This policy was subsequently reversed when it was recognized that without the shade, the coffee became more susceptible to disease and more dependent on artificial aid.

Other examples of highly managed tree-production systems can be found elsewhere in the SADCC countries. In a number of moist coastal regions, cashew trees and coconut palms are extensively grown as cash crops. In other parts, citrus orchards are common.

In cases such as these, farmers do not need to be told how to grow trees. On the contrary, they may have a lot to teach the forester, who has normally been trained only in the techniques of managing a few well-known timber species. Local knowledge is often much broader based. It covers a wider variety of species and, usually, a wider range of tree products. Rather than harvesting for a single purpose, the people know how to obtain many different uses from the same tree.

Medium-Intensity Tree-Management Systems

More widespread are what can be classified as medium-intensity tree-management systems. In these, farmers are more selective in the trees they grow. They plant fewer, and generally rely on natural woodlands for a substantial part of their wood needs. The importance of trees in these farming systems, however, should not be underestimated. They often play a critical and highly valued role.

Growing fruit and shade trees around the homestead is probably the most common form of tree cultivation. This can be observed throughout the SADCC region, almost everywhere that settled agriculture is practised. When families move into a new area, planting a few trees around the home is often one of the first things they do. This tradition often continues in the urban area.

Mango trees are one of the most popular species for planting around the house. Their dense foliage provides ample shade, and their heavy crop of fruit is an important source of food during several months of the year. Jacaranda is widely grown as an ornamental tree in some areas. Other trees are valued for the timber they produce, or for their termite-resistant wood.

Using trees or shrubs as live hedges is a traditional practice in many SADCC countries. The species used depends on the area. Around Zomba, in Malawi, *Thavitia spp.* are the most common. Trees are raised by transplanting naturally occurring seedlings. They grow fast and can be pruned to provide wood for the farm.

Around Lake Malawi, on the other hand, farmers prefer *Cassia siamea*, which they grow by direct seeding. In other areas, mauritius thorn (*Caesalpinia decapetala*) is popular, while in drier regions such as Swaziland, live hedges made of sisal (*Agave americana*) and *Euphorbia tirucalli* are more common.

Some farmers devote part of their land to individual woodlots. In Zimbabwe and Malawi, eucalyptus is the most popular species for this purpose. In most cases, woodlots are planted primarily for poles and construction timber, with firewood being a by-product. In Swaziland, plantations of black wattle, originally introduced for tannin extraction, are also quite a common sight.

The total number of trees grown on farms varies a lot from region to region. Within individual villages there are also significant variations in tree-growing practices. Some farmers are much more interested in trees than others. They have a greater knowledge of the properties of different trees, of the best techniques for propagating them, and of the most effective ways of harvesting them to ensure maximum sustainable yields.

These variations mean that making generalizations is often dangerous. Underestimating the extent of local tree-growing knowledge has frequently been a fault in the past. This has meant that many tree-growing programmes have missed out on important opportunities to tap into this knowledge and build on the traditions that already exist.

But it is equally true that while most farmers know the basics of how to grow a few well-known species, not everyone is an expert. Farmers could often learn a lot from one another, and could pick up useful tips by observing what villagers do in other areas. Finding ways to share and exchange this traditional knowledge is an area of great untapped potential (see Box 3.1: Learning from your Neighbour).

Low-Intensity Tree-Management Systems

The last category is low-intensity tree-management systems in which the deliberate planting of trees is rare, and management of trees is mainly restricted to selective harvesting of natural vegetation. These systems are found particularly in the drier zones where populations are sparse and agriculture is dominated by livestock. Here the balance between people, animals and trees is quite different

Box 3.1 Learning from your Neighbour

Outside experts are not always needed to spread knowledge on tree-growing techniques. Farmers could often learn a lot from each other. This was well-illustrated in one village visited in Malawi, about 20 kilometres south of Zomba. Here, two farmers, separated only by a stream, were using quite different methods for propagating eucalyptus seedlings.

On one side of the stream was Joel Banda's farm. Recently married, he said he had a lot of problems finding good-quality timber for building his house. He was forced to use inferior wood, both for the roof beams and for the supports at the front. To provide building wood for the future, he had decided to establish his own eucalyptus woodlot, using seedlings purchased from a forestry department nursery a 10 kilometre walk from the village. His fields were being prepared for planting beans and sugar-cane as soon as the rains came. Hand-ploughing with a hoe was followed by gathering the weeds in small piles. After drying in the sun, the piles were burned and the ashes used as a seed-bed for raising tomatoes and pumpkins. Every day these small vegetable nurseries were watered from the nearby stream.

Just across the stream were the fields of Elison Mwale. He used the same technique for preparing land but in the seed-bed created from the ashes, grew not just vegetables but eucalpytus seedlings as well. Directly after seeding, the bed was covered with dry grass to shield the young plants from sun and rain. They were tended and watered in the same way, and once they were strong enough, were transplanted to the compound around his house, to protect them against theft and animals.

Mwale learned the nursery technique from an older man in the village. Eucalyptus seed is collected by climbing a few of the large trees in the valley that were planted some 15 years ago by his uncle. After the seed pods are collected, they are dried for two weeks in an old plastic bag. After this period the seed is separated out and mixed with some yellow sand from the nearby road. Sand-mixing is important, he says, otherwise the small seeds will be sown too densely in the bed. By raising seedlings himself, Mwale saves money as well as a long journey to the forest-department nursery. The success of his methods is clear from the large number of trees growing around his home. Others in the village could learn from his example and his neighbours could benefit substantially – without the need for expensive inputs.

from that in the more densely populated areas. Growing trees is much more difficult, both for environmental reasons and because of the problem of protecting trees from grazing animals.

Instead, the system is based almost entirely on the regenerative capacity of the natural vegetation. Nevertheless, trees play a vital role in livestock production, providing fodder (including important nutritional supplements), browse, shade, and rubbing poles for pest control.

Low-intensity tree-management systems can be found in all the SADCC countries. In Botswana, Lesotho, Zimbabwe and Swaziland, it is the most widespread system in operation – at least in terms of land-area covered. In other countries, it is restricted mainly to the drier regions such as Namibie Province in Angola.

Just because management of trees is passive, however, does not mean that people are ignorant about trees. In fact, knowledge of the different species is often greater than in other areas where farmers are more active in growing their own trees.

Local knowledge

Since they depend on natural woodlands for a wide range of essential products, people have learned over generations which species are useful for what purposes. Details vary but beside firewood, the list of products obtained includes building timber; wood for kraal fences, tools, transport and construction (boats, scotchcarts, sledges, etc.); edible leaves, pods, nuts and fruit; honey; natural fibres; fodder; medicines; utensils; and a whole range of other items. The cutting of trees is rarely indiscriminate and people often have considerable knowledge about how to obtain the most from the woodland around them.

Under the *chitimene* system in Zambia, land was traditionally left fallow for at least 25 years. Recently, however, the system has come under increasing stress owing to the expanding population and the reduction in the area available for farming.

Fallow periods have fallen to only 10 years in places, and some farmers have begun to establish permanent gardens. The long-term impact of this is still unsure, but the likelihood is that crop yields will gradually decline as the soil becomes more impoverished and the overall productivity of the system is eroded.

Similar problems are emerging in other parts of the SADCC region. As human and livestock populations grow, traditional systems of low-intensity tree-management are showing signs of breaking down.

It is important to recognize that even though traditional systems may become outmoded as they are overtaken by events, the knowledge and experience on which they are based is still a valuable resource. Local people frequently know far more about how to manage their surrounding environment than any outside expert. To discard this knowledge as irrelevant to modern conditions is both arrogant and short-sighted. With external inputs and support, it will often provide vital pointers towards long-term solutions that are both sustainable and locally viable.

REFERENCES

1. P. Harrison, *The Greening of Africa* (London: Paladin Grafton Books, 1987).
2. World Commission on Environment and Development, *Our Common Future* (Oxford: Oxford University Press, 1987), p. 43.
3. P. Blaikie and H. Brookfield, *Land Degradation and Society* (London: Methuen, 1987), p. 7.
4. Ibid., pp. 7 and 8.
5. R.W. Franke, "Power, class and traditional knowledge in Sahel food production", in I.L. Markovitz (ed.), *Studies in Power and Class in Africa* (Oxford: Oxford University Press, 1987).
6. Ibid., pp. 259 and 260.
7. E.P. Thompson, *Whigs and Hunters. The Origin of the Black Act* (Harmondsworth: Penguin Books, 1977).
8. W. Beinhart, "Soil erosion, conservation and ideas about development: a Southern African exploration, 1900–1960", *Journal of Southern African Studies*, Vol. II, No. 1, 1984, p. 84.
9. H. Kjekshus, *Ecology Control and Economic Development in East African History* (London: Heinemann, 1977), p. 184.
10. L. Vail, "The political economy of east–central Africa", in D. Birmingham and P.M. Martin (eds.), *History of Central Africa* (London: Longman, 1983), pp. 200–250.
11. T.O. Ranger, *The Agricultural History of Zambia* (Lusaka: National Education Company of Zambia, 1971).
12. See for evidence of this in the case of Mozambique. R. First, *Black Gold* (Brighton: Harvester, 1983).
13. E. Boserup, *The Conditions of Agricultural Growth* (London: Faber, 1965).
14. W. Brusse, *Bericht uber eine im Auftrage des Kaiserlichen Gouvernements von*

Deutsch–Ostafrika ausgefuhrte Forschungereise nach dem sudlichen Teile dieser Kolonie (Berlin: Mittler & Sohn, 1902), quoted in H. Kjekshus, *Ecology Control and Economic Development in East African History* (London: Heinemann, 1977), pp. 38 and 40.

15. G. Clarence-Smith and R. Moorson, "Underdevelopment and Class Formation in Ovamboland, 1844–1917", in R. Palmer and N. Parsons (eds), *The Roots of Rural Poverty in Central and Southern Africa* (London: Heinemann Educational Books, 1977), p. 98.

16. Zimbabwe Energy Accounting Project, *Household Rural Energy Study* (Harare; Beijer Institute and Ministry of Energy and Water Resources and Development, 1984). These findings were based upon a systematic sub-sample (N = 200) of the 1,000 household rural energy study. Seventy per cent (n = 142) of the households of the systematic sub-sample indicated that they had planted trees and the remainder either indicated that they did not plant trees or did not respond. Dr Yemi Katerere analysed the results of this survey.

17. FAO, *Tree Growing by Rural People* (Rome: FAO Forestry Paper 64, 1985), pp. 13–14.

4. Tackling the Problem on the Ground

In this chapter we look at concrete experiences within the SADCC region. These help to show ways of tackling the problem in different locations and circumstances. The case studies demonstrate that there is no simplistic formula for success. The complex local circumstances, both environmental and socio-economic, have to be taken into account if a successful policy is to be developed. Each of the case studies drawn from within the SADCC region, is intended to demonstrate how a new approach to the problem is already taking place and can also be improved upon.

Some of the case studies demonstrate wholly indigenous initiatives in woody-biomass management while others show how outside agencies can encourage a participatory approach which builds upon existing local tree knowledge. A major unifying theme is the need to improve land-use management practices which can boost production and improve the woody-biomass component of the production systems.

The sites were chosen following the identification of major fuelwood problem areas within the SADCC region. This identification was made on the basis of a remote-sensing exercise integrated with ground verification and local information.[1] Each of the sites chosen provides valuable lessons which illustrate the validity of the approach that we are proposing.

The first study examines a high-potential arable-farming region in central Malawi where there is evidence of medium- to high-intensity tree-management by smallholder-farmers. This represents a successful case of planners encouraging and facilitating smallholder tree-production by expanding the sphere of activities of the forestry department. However, the problems facing the large number of smallholder-farmers who are experiencing a major land shortage have still to be resolved. This case also highlights the problems associated with tobacco production on estates, which places a heavy strain on fuelwood resources and for which a number of remedial measures are proposed.

Next we look at two locations in the south-east of Zimbabwe which are semi-arid, mixed-farming areas under communal tenure, with medium to low tree-management systems. Gutu has a major programme based on ensuring environmental regeneration and sustainability with improved land-use management practices incorporating an array of agroforestry interventions. This externally initiated scheme is the first large-scale attempt within the SADCC region to incorporate a woody-biomass component in the improvement of a production system. The Mwenezi scheme, on the other hand, represents an indigenous initiative for improved land-use management, including rotational grazing of livestock in indigenous woodland.

Shinyanga, the next case study, is one of the most desertified regions of Tanzania and water shortage makes the extensive development of tree nurseries unlikely. Yet local people are having some success with tree cuttings and direct seeding. Where grazing and burning are being controlled by the community itself, there is evidence of rapid natural woody-biomass regeneration. The Swaziland case points to a similar lesson but shows, too, how the prospects of tree-growing for cash (in this case, the production of wattle for tannin) can indirectly result in an important fuelwood resource. Finally, the managers of the forest reserves in the Dedza-Ntcheu highlands of Malawi, surrounded by a hinterland scarce in fuelwood, have shown great innovation in devising an array of methods for the improved management of indigenous woodland and forest for fuelwood resource-sharing.

All of these cases demonstrate the importance of identifying and evaluating the land-use and woody-biomass management systems. Following this, the existing constraints can be appraised and the potential for improving the productivity of the woody-biomass management system can be assessed.

LILONGWE PLAIN, MALAWI

The Lilongwe Plain is an area of high wood demand and scarce indigenous woodland. All the evidence suggests that there has been rapid deforestation on the plain over the last 15 to 20 years. Now, the only pockets of indigenous woodland remaining are the traditional clan graveyards which custom dictates must be left untouched. The remaining indigenous woodland is to be found around the periphery of the plain, mainly in the form of forest reserves, and this is the major source of fuelwood for the capital, Lilongwe. This situation has resulted from a population increase and a major expansion of arable production, cash-cropping in particular. There are two major forms of agricultural-production system: estates on leasehold and freehold land; and smallholder production on land occupied under customary law.

There is a deficit of arable land within Malawi as a result of growing population pressure and the current division of land by different systems of tenure and use.[2] After independence, the remaining uncultivated arable land was rapidly acquired by an expanding estate sector. The percentage of cultivated land occupied by the estates increased from 1 per cent at independence to 16·5 per cent by 1984.[3] Between the late 1960s and the early 1980s the number of smallholdings increased by 22 per cent but the average size of landholding declined by a quarter to 1·16 hectares.[4] The migration of labour from the smallholder to the estate sector increased as a result, reinforcing the crucial role played by female labour in smallholder production.

The expansion of the estate sector, based mainly on tobacco production in the centre of the country, led to rapid deforestation. Tobacco estates currently use 23 per cent of all wood energy consumed in Malawi.[5] In the smallholder sector, population pressure, limited land availability and land clearance for agricultural production have placed severe stress on the woody-biomass supply in a number of places. Experience from both sectors highlights the need to seriously consider the impact of agricultural strategies on woody-biomass supply.

Low wages and the low prices smallholders can charge for their produce have tended to limit any transition up the domestic energy ladder, in particular given the government's full-cost pricing policy for commercial fuels. As we will show, there have been important energy-planning innovations which will help ameliorate the fuel-wood problem for some if not all of the peoples affected by these constraints. Although we have earlier urged caution over such predictions, the matter is urgent. One estimate shows that natural reserves of fuelwood nationwide could be exhausted by 1990, given the projected overall increase of demand of 3·5 per cent per annum.[6]

Tobacco estates are required to leave 10 per cent of the land forested but this rarely happens and considerable legislative revision and proper implementation of the law is imperative. The provision of wood clearly varies from estate to estate. In the large estates of General Farming and Press Farming, a senior forester has been employed since 1979 to undertake an extensive planting programme. However, this remains rare. The cheapness of the wood available elsewhere, either from state forests or customary land, acts as a disincentive to estates growing their own plantations. Smallholders frequently allow the free use of wood on the customary land, in exchange for the land clearance undertaken for them by the estates. According to the Energy Studies Unit in Malawi, the current price of wood will have to more than double in order to make it profitable for estates to plant trees themselves.[7] As this rise in costs is envisaged anyway by the Forestry Department, there may be a greater incentive in the future for estates to be more self-reliant in the production of wood.

Little forestry extension-work has been carried out within the estates but greater attention will be paid to this in the future. *The Tobacco Sector Study* proposed a licensing system for flue-cured Virginia curing-barn furnaces to ensure both the planting of adequate fuelwood reserves and the more efficient utilization of these resources.[8] More specifically, alterations in barn structure, furnace design, flues, chimneys and ventilators can improve efficiency in wood consumption. The *Flue-Cured Tobacco Energy Use Survey* carried out by the Energy Studies Unit, makes a series of practical recommendations.[9] These include minor improvements to existing barn structures; estate extension; the training of barn operators; and greater government encouragement to estate tree-planting and conversion to other energy sources, such as coal.

Extensive land clearance, fueled by a growing population and a major World Bank agriculture project, markedly reduced the available indigenous woody biomass. A major rural-planning initiative, the Wood Energy Programme, was set up in 1980 funded by the World Bank. Phase I ran from 1980 to 1985. The central thrust of the programme is farmer-based afforestation and there is a clear move towards a true farm-systems approach to farmer tree-production, and increasing recognition of the traditional skills of farmers in tree husbandry. A wide network of nurseries was established to sell seedlings to farmers. In this area, we find a medium to high tree-management system.

The Lilongwe Plain experience exemplifies a planning intervention targeted on the individual farmer in a predominantly arable production-system based on intensive farming. It aimed to provide fairly short-term returns to the individual farmer through pole production from woodlots with the appropriate emphasis on mono species. But it also included the encouragement of multi-purpose species for agroforestry use. Around the periphery of the plain there was also conservation intervention with the protection of natural woodland. It is necessary, however, to assess the programme's potential for the different strata within rural society.

The richer smallholder-producers with initial capital, larger landholdings, greater labour availability and access to institutions and/or transport, appear to have shown their willingness and ability to convert part of the arable land to woodlots for construction-pole marketed production as prices have risen. The economics of fuelwood production do not as yet make this an attractive commercial venture. At the same time, these farmers are meeting their own needs for fuelwood and construction timber. The Wood Energy Division's strategy is to continue to encourage pole production, even though a surplus may soon be reached, as this may enable the traditional preference of indigenous wood for fuel to be breached, with a greater acceptance of exotics.

For those with the largest smallholder land-holdings there is a tendency towards monocropping of poles on the land allocated. Those with slightly less land tend to intercrop with maize, or another crop, in the first couple of years to offset the loss of marketed maize returns. With proper management, returns on poles can be obtained within five years. One of the major side effects of intercropping is that it ensures that the woodlot will be weeded. The woodlot serves as both investment and insurance, making optimal use of the farm resources available as it is usually located on the poorer soils within the farm.

Trees can be an important crop when the life cycle of the smallholder family is considered. As the farmer grows older and children leave home, there is less labour available for maize production, and pole production with its lower labour requirements, becomes increasingly important. While the area committed to pole production will vary from farm to farm, site visits indicated that at present it is unlikely to exceed 20 per cent of the available area and is often much less. Farmers with well-established pole plantations indicated that the return is higher than that for maize which has a fixed price ceiling.

For the middle strata of smallholder-producers with more limited capital, labour, available land and access to institutions, there is less pole-growing for marketed production. Some plant small woodlots to meet domestic needs for construction with the residual benefit of fuel. Surplus production tends to be minimal, although as the price of poles rises, it is possible to envisage some potential here for growth.

For the poor strata of smallholder-producers, a major constraint is the extremely limited size of their landholding and possibly also a labour constraint, as the men are obliged to work away from home. There is also limited capital and access to institutions. These families may be late-comers or may have suffered from land fragmentation through inheritance. Here there is no land physically available for woodlots. It is a measure of the active awareness of the problem that a widespread number of responses are in evidence. In particular, there are fruit and shade trees around the homestead; and the use of live fences (with sisal as sapling protection) for poles, fodder, fuelwood, and the demarcation of vegetable gardens and boundaries. Indeed, such practices are to be found among all farmers. Mangoes are left in the fields and provide a valuable and convenient source of food in the hungry period of January–February, prior to the harvest. It is a measure of the wood-shortage problem on the plain that there is evidence of the partial and total felling of mangoes. Domestic energy needs are being met with increased scavenging for fuel, which is labour-intensive, and a greater use of crop residues, particularly corn stalks.

It has proved much more difficult for governments to intervene positively to help tackle the energy needs of this poor stratum of farmers, who comprise the bulk of the population. They lack sufficient land and capital resources to await a delayed return of five years in wood production for land investment.

A current problem facing the Forestry Department in Malawi is a result of its success in encouraging greater pole-production over a wide area. Some farmers are now complaining that they find it difficult to market their poles. Hence the Forestry Department is considering how to extend the range of support that it provides, perhaps by publicizing the availability of poles and facilitating the meeting of purchasers and suppliers. One possibility would be the provision of a revolving fund for the Department of Forestry to purchase the available poles, which it would then market. Small mobile saw-mills could also be employed. All this implies a broadening of extension work along the chain from producer to consumer.

There is plenty of evidence available on the ground to demonstrate the extent of people's responses to the woody-biomass scarcity already occurring. There is evidence, for example, of Gmelina and sisal fencing combined for erosion control, construction, and fuelwood purposes. Agroforestry practices are there for all to see. But with the exception of the graveyards, customary woody-biomass management practices are being eroded. The indigenous plots of woodland that chiefs preserved for fuel to be used at festivals, and for the old and infirm, have disappeared as the pressure on the land increases with the growing population.

Future interventions need to focus upon the actual constraints that the poorer majority of the rural population have to face. In particular, research and extension activities need to concentrate on multi-purpose trees suitable for integration into the production systems of smallholder-farmers with limited land and other resources. Given the central and growing role of women in smallholder-production and the increased time that the men spend working away from the farm, the support given to women farmers by female extension agents is a high priority. In order to implement these and other necessary measures, important changes in the training of Forestry Department and Ministry of Agriculture personnel is urgently required. The restricted areas of land make increasing the intensity and diversity of on-farm land-management for production the central concern.

Let us now turn to a less well-endowed agricultural area where mixed farming predominates. Here we can see examples of both an indigenously and an externally initiated project with important woody-biomass implications.

GUTU, MASVINGO PROVINCE, ZIMBABWE

In many parts of the communal areas of Gutu district, fuelwood provision is a serious problem with a high demand being made on a poor woody-biomass resource base. Land degradation is high and, arable cultivation is encroaching on available pasture land. This produces a long-term downward spiral of diminishing returns, promoting the disintegration of the traditional mixed arable and pastoral production system. The woody-biomass resource continues to diminish, given indiscriminate deforestation under the priority of cultivation. Demands on women's labour-time for fuelwood collection are increasing along with a tendency towards the commercialization of firewood.

The origins of the problem lie in the colonial period. Settlers alienated African farmers from the best land, creating labour reserves in pockets of communal tenure. These Communal Areas proved increasingly inadequate to support a growing population which was forced to rely more and more on male migrant labour. In Gutu District, pockets of tree-less and shrub-less barren land lie next to the well-wooded pasture of the settler-farmers.

There is considerable local diversity in woody-biomass availability. Figure 4.1 shows people's perceptions of the availability of fuelwood in four different locations within Gutu District.[10] Considerable indigenous woodland remains in the hilly agricultural marginal area of Kubiku. Here, fuelwood is not a serious problem. In Denhere, 88 per cent of the land is already cultivated and all indigenous woodland has been cleared. However, people's perceptions are of a moderate availability because they illegally collect fuelwood from the neighbouring commercial farm land. An informal and *ad hoc* situation of resource-sharing mitigates the problem here. Nerupiri and Central Gutu experience a severe scarcity of fuelwood with the pressure of a high population and extensive deforestation. Collecting wood requires long, time-consuming trips. Central Gutu is situated far from any indigenous woodland on commercial farms. Here, fuelwood is one of the most pressing problems that people have to face.

At the micro level, fuelwood shortages depend on access, which in Gutu is influenced by a number of factors:

1. Individual or family claims to remaining supplies located near homesteads, which are defended against "outsiders".
2. Transport: the ownership or possibility of hiring a scotchcart increases access while reducing the overall time spent collecting fuelwood.
3. A co-operative husband: the operation of a scotchcart is traditionally a man's task and firewood collection is a woman's task.
4. The number of daughters, daughters-in-law or young sons present to help with fuelwood collection.
5. Health and old age prevent some from making longer trips to remaining stocks.

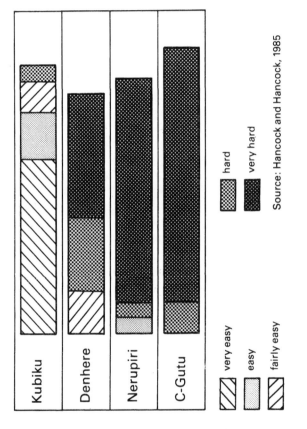

Figure 4.1 Gutu people's view of their access to firewood

6. Tools, e.g. an axe or saw for cutting wood (some poor households had no axe).
7. Purchasing power (the ability to buy firewood).
8. Proximity to either large-scale or small-scale commercial farming areas where wood is generally collected on a basis of informal resource-sharing.

This example indicates the need to move beyond the identification of a problem at district level alone. Instead, the specific people and places where the problem is most keenly felt must be located.

In areas like Gutu, with a high population density, poor soils and low rainfall, the risks of crop failure are high and yields tend to be low. There is a rapid deterioration in the land with erosion and declining fertility. In these circumstances, the ecological balance must be restored and a suitable, environmentally sound series of farming practices needs to be established.

Within the Gutu District, an integrated approach to rural development is being established with the Co-ordinated Agricultural and Rural Development (CARD) programme. This is a systems approach to development in rural areas which co-ordinates institutional efforts and sectoral needs. The aim is to guarantee a viable subsistence through ecologically sound farming practices. Such an approach necessarily incorporates improved woody-biomass management practices. In terms of planning choices, the CARD programme represents a farmer-based, broad focus, long-term planning intervention incorporating multi-use tree species.

Given the high population density of the district (46 persons per km^2) and constraints in both the area and quality of land available, there is a serious deterioration of the resource base. High-input farming practices run the risk of increasing debt, given poor soils and uncertain rainfall, or the exclusion of certain farmers from loans to purchase inputs if they are not considered good prospects for credit.

With growing population pressure on the limited resource base in the communal areas of Gutu, arable land has encroached upon pastoral land and only 30 per cent of the land area remains as grazing. A project aim is to increase the intensity of crop production in order to reduce the expansion of arable land. At the same time, the scheme intends to improve cattle production with better grazing schemes, improved herd quality, cash earnings, and manure production for increased fertilizer use.

Intercropping and agroforestry practices will be introduced to increase the nitrogen content of the soil and reduce the need for commercial inputs. This represents the first major attempt within the SADCC region at introducing a systems approach in an integrated rural development project which also includes a woody-biomass production component. This strategy offers alternatives to the progressive farmer high-input philosophy which has dominated the extension programme to date, often ignoring the resulting cost to both farmer and country, given a progressive ecological decline.

The problem is that such intercropping techniques improve soil quality in the long term while possibly, but not inevitably, carrying a short-term cost in terms of a small diminishment in yield. It is possible, however, to guarantee that any shortfall to the farmer will be made up if required, until the system is established.

This will help to break down the initial resistance of local farmers who cannot sustain short-term risk against long-term gain. The enhancement of ecologically sustainable production systems is part of the process of addressing the causes rather than the symptoms of rural poverty.

Such a strategy provides potential advantages both in terms of national food security and in meeting the fuelwood shortages. Hence an integrated approach may allow multiple and interactive problem areas to be tackled. The increased intensity of arable production allows less deforestation for land clearance and agroforestry interventions in the new arable and pastoral management practices, all of which help tackle the fuelwood problem.

The potential for intervention in this situation is detailed in Table 4.1. All or a combination of interventions can be chosen to suit the particular circumstances of a given area, and many are in the CARD project.

Given the extremity of the situation in some areas, other measures can be taken. In circumstances where this is politically possible, the resettlement of families can serve as an important temporary solution. However, high population growth may limit its long-term impact on communal areas. Given the dependence of many rural households on cash from migrant labour, it is important to consider ways of increasing employment opportunities. This allows access to fuelwood as a bought commodity and commercial fuels, in addition to satisfying other basic household needs.

People living in parts of the communal areas where fuelwood is scarce, adjacent to commercial farms with extensive woody-biomass resources, are already practising resource-sharing, usually in the form of poaching. A wide array of opportunities exist for planning this resource-sharing in a manner that does not lead to environmental degradation and conflict. We will return to this issue again later in this chapter.

The Gutu experience in the CARD programmes shows how improved arable and pastoral management practices involving the provision of woody biomass can help address the fuelwood problem.

Table 4.1 Options for Increasing Woody-Biomass Production and Management on Communal Land

1. Resettlement to ease the pressure on the land resource base.

2. New land-management practices – demarcation of settlement, arable and pasture areas.

2.1 Pasture*	– paddocking using live fencing for fodder, fuel and poles
	– preserving indigenous wood in paddock areas (shade and erosion control)
	– planting fodder trees scattered or in lines for ease of control
2.2 Arable	– intercropping for nitrogen fixation and woodfuel off-take
	– live fencing for wind breaks, erosion control, fodder, soil fertility
	– intensification of crop production by new land-management practices to reduce deforestation for arable production
2.3 Settlement*	– trees for fruit (can be multi-purpose, trees for shade, trees for poles, and all can have a fuelwood off-take)
3. Nurseries	– government run
	– schools, councils, co-operatives
	– individual farmers
4. Woodlots	– institutions, schools, councils
	– community
	– individual farmers
5. Orchards	– fruit, nuts (fuelwood, off-take)

---

* Indicates measures that can be used by individual homesteads based on existing land-use models or in improved management schemes.

These can be supplemented by rural afforestation and woodlot programmes. *The emphasis here should be on extension work rather than the production and management of woodlots by the government.* This is the approach increasingly being adopted in the second phase of Zimbabwe's Rural Afforestation Project (RAP). This provides an example of a low-cost, decentralized, informal planning initiative, with a broad focus on multi-use tree species. The long-term aim is disseminating and expanding tree-growing at both a domestic and a communal level.

Separate from the CARD project, about 80 schools in the Gutu District in 1985, were being encouraged to develop nurseries at a cost of less than 1 cent per tree as opposed to 10 cents per tree in the RAP nurseries. At a total cost of about Z$ 100, it is possible to establish a 5,000–10,000 tree nursery in a school. The best school-nursery in the District, at Chipangane, produced 20,000 trees in the 1985 season. In just one part of the District, nine Village Development Committee nurseries were also established.

In Gutu, we have the valuable experience of an *externally* initiated integrated project with an important woody-biomass management component. Next, we examine an important *indigenous* planning initiative taking place in another part of the same province.

MWENEZI, MASVINGO PROVINCE, ZIMBABWE

This is a semi-arid area of medium to high population density and extensive woody-biomass growth. In Mwenezi, there is a heightened local perception of the problem of increasing environmental degradation and the attendant poverty of the population. This has created an openness towards tackling the problem of improved land-management. There is also the political will present within the community and a readiness on the part of the government and donors to support such local initiatives.

Rainfall is low and the area experiences periodic droughts. On average, there is only one good year of rainfall in every five. The district is a part of the labour-reserve economy dependent upon male migrant-labour earnings and peasant agricultural production of livestock and grains at subsistence level.

The pressure of a high population on a poor natural-resource base has led to increasing land degradation. Returns from arable and pastoral production are low and there is a growing vulnerability to drought. Water is a limiting factor. Boreholes frequently dry up, as does the Mwenezi River. Cattle are a vital part of the farming system but over 40 per cent of the population do not own them. As a consequence, considerable social differentiation exists within the local community. The extended period of drought between 1982 and 1984 severely affected the population, diminishing crop production and depleting herd sizes.

Many of the men are engaged in wage labour in the towns, mines, large agro-industrial schemes operating in the south of the Lowveld, and to a lesser extent in local growth areas. Population growth and limited wage-labour opportunities have created high unemployment and attendant poverty made worse by the inherent weakness of the subsistence-farming base. Extensive drought relief was received in the recent past. The limited opportunities for wage-earning and minimal-surplus production place severe constraints upon the ability of households to purchase commercial fuels. The heavy reliance upon fuelwood is compounded by traditional preference and fire-management practices.

A new system of land-use-management was introduced by local initiatives in 1982 and is currently operating in three out of the twenty-four wards in Matibi I, one of the two communal areas within the district. Each ward contains approximately 600 households (or 3,000 people). It has an average of one livestock unit (LU) for 3 hectares (ha), whereas the recommended stocking-level under managed conditions is 1 LU per 10 ha. There is an obvious need, to implement improved management practices and arrest the land degradation that is occurring. This involves increasing the number of cattle for sale and improving pasture management.

The new system of land-use-management involves the separation of settlement, arable and pastoral areas. On one side of the settlement is the demarcated arable land, with pasture area on the other. Within the pasture area, a series of paddocks are marked out to permit rotational grazing. This system of "veld management" was first tried out within the province in the early 1970s.[11]

The grassroots initiative taken by active leaders within the local community provided an important ingredient for the success of the new scheme. Of the three wards currently involved in the scheme, Ward 14 is the most advanced. In this ward, seven paddocks have been fenced off within the indigenous woodland area and a 21-day rotation established for the 156 cattle currently involved. Preserving the trees helps to maintain the ecological base of the pasture-improvement scheme. Currently, wire fencing is used, purchased by an EEC grant. However, its high cost and the threat of sanctions by the Republic of South Africa (the source of much of Zimbabwe's wire fencing), presents opportunities for the development of live fences. If live fences are established, the existing wire fencing can be used to establish further paddocking schemes. Live fencing will provide multiple additional end-uses including fodder, poles and fuelwood.

Attempts to establish gum-tree woodlots have failed, indicating the need for further research on suitable species for semi-arid zones. In addition, macademia and cashew tree orchards have been introduced. These provide the opportunity for earning cash and a residual supply of fuelwood. There are also measures being taken towards gulley control, including the growth of sisal which also supplies fuelwood.

Paddocking allows a major saving of labour-time, as the burden of herding is reduced. This allows more time for gathering fuelwood and fire-management, in addition to adult literacy, children's education, arable production and domestic labour.

The short-term benefit of this community-integrated initiative is already apparent. On average, cattle that previously fetched Z$100–150 are now sold for Z$400–450, providing greater incentives for farmers to market cattle. Increased cash earnings permit some fuelswitching, in this case to kerosene, for lighting and the fast boiling of water. The greatest constraint for fuelswitching is the uncertain reliability of supplies.

The Mwenezi case study offers numerous opportunities for the integration of woody-biomass production in the mixed crop and grazing scheme. In addition, there are opportunities for the introduction of improved management systems at all farming levels. These options have yet to be fully explored and implemented in Mwenezi. It is important that this community-initiated effort be developed further and that the use of local inputs be encouraged and strengthened.

The opportunity exists within the grazing paddocks for improved pasture as well as management of indigenous woodland. The planting of trees for fodder can be introduced and some can be planted as boundary fences. The practice of rotation grazing will reduce the problem of browsing. Legumes can also improve the quality of pasture and greater use of indigenous species as fodder could also be encouraged. The pods of certain acacias such as *Acacia*

Sibera are excellent fodder when dry. *Brachystegia sp* produces multiple coppices which can reach a height of 1·5 metres in about two years. The coppices are excellent browsing for cattle when grass is in short supply but this will require careful control. The growth characteristics of these indigenous trees makes them a potential source of fuelsticks.

Placing cattle in paddocks increases the potential for introducing woody biomass (including fodder) to agricultural land. The chances of a successful agroforestry type programme are therefore enhanced.

The Mwenezi experience is important in that it discards the long-accepted notion that the communal tenure system (e.g. communal grazing) is the greatest constraint to the implementation of meaningful intervention strategies. As smallholder-farmers are confronted with impossible odds for survival, they can become more open to trying out new management techniques that help address their priority concerns. The experience is also valuable for it shows what measures can be taken to avoid rapid deforestation in areas not yet facing a fuelwood deficit. Prevention is better than a cure.

SHINYANGA, TANZANIA

This is an area of high demand on a poor woody-biomass resource base. Shinyanga is one of the most desertified regions of Tanzania. Farmers spread their risks by having no uniform farming system. Capitalizing on diverse local conditions, they engaged in subsistence and cotton production, with cattle as an investment. Between 1960 and the late 1970s, population density in the area increased by 34 per cent and the index of agricultural activity increased by 73 per cent while the "wooded area" index decreased by 73 per cent.[12] Population increases led to the establishment of new villages, reducing access to land and a risk-minimizing diversity of soil conditions. The capacity for biomass reproduction is rapidly being lost with topsoil eroding at the rate of 1 to 2 mm a year, and double this rate around villages and stream-beds. People are increasingly relying on residues for fuel as fuelwood is becoming a commodity. The water supply is diminishing, grazing and arable land deteriorate, and people can no longer find wood to make implements and have to buy them. Even health is affected as suitable timber is not available for building latrines.

Tree-planting has been encouraged and farmers now plant an

average of three trees per hectare each year. This is one-tenth of the rate needed to redress the calculated fuelwood deficit. Communal planting has been tried, without great success. The Forest Division plans to provide over 100 nurseries (one per ward) to increase seedling production of agroforestry species for individual planting. This traditional approach is unlikely to succeed however, because of

1. Severe water shortages in most villages, which make seedling care very difficult.
2. Impassable roads and expensive transport to nurseries at water points.
3. Lack of protection of young trees from cattle which graze freely on fallow and common land.

Land around homesteads is fenced off from cattle and supports trees – virtually the only ones visible. But this is typically less than one acre per family and the trees are used for shade and poles as well as fuel. Common grazing constitutes 95 per cent of all land.

Other possibilities for tackling the problem do exist. Farmers have invented several ways of propagating woody biomass other than growing from seedlings. Live fences of euphorbia and cassava-like species are universally grown from cuttings, positioned along paths and near houses. Both are used for fuel. At least 10 species of "real" trees are also grown from cuttings including acacia, pomegranate, commiphora, ficus and gliricidia. Guava trees are grown from root cuttings. *Direct seeding* is practised for even more species, including three types of acacia, leucaena, mango, custard apple, tamarind and cashew. Some of these are nitrogen-fixing and enhance the soil's fertility. Much more could be done to spread these low-cost techniques. Foresters are trained in them but tend to forget them because of the emphasis on nurseries and seedling distribution.

Shinyanga was deforested by clearance for tsetse-fly control, followed by heavy grazing and the practice of early burning. This prevents the regrowth of natural acacia. The carrying capacity of one livestock unit per four hectares is exceeded by a factor of four over the region as a whole. Since large cattle herds are maintained to spread risks during drought periods, de-stocking is not at present a feasible solution. However, while the Shinyanga plains are bare of trees, there is a 40–50 hectare Forest Division reserve from which cattle have been excluded since 1981. The rate of acacia regrowth in this area, which had adequate water, has been phenomenal. After five years, the canopy is about five metres high. Given the vast areas of grazing land, *natural regeneration* is the key. The question is how to keep selected areas free from cattle and burning. Again, several spontaneous approaches have emerged. Some villages have established thorn hedges to protect young acacia saplings; some have forbidden grazing on limited tracts of land (with heavy fines as penalties); some have set aside special areas for the gathering of fuelwood, which is banned everywhere else. These initiatives need to be supported, built up and spread in every possible way. Village councils are well able to impose community sanctions provided there are incentives to farmers, such as veterinary services to improve cattle herds, or better water supplies.

SWAZI NATION LAND, SWAZILAND

Swazi Nation Land (SNL) is under communal tenure. Typically, the situation is as follows:

Grazing Land. Is generally treeless with patches of wattle of various sizes, some clearly planted, others self-seeded in valleys. Less frequently, thickets of thorn and wild guava are found on poor land. On the steep and higher hills there is some thin woodland, scrub and aloes.

Arable Fields. There are isolated protected trees in or beside the fields. As a form of erosion control, grass strips are left between the fields and there is some broken fencing of aloes, cacti, sisal and live branches.

Homestead/Yard. Fruit trees and a variety of live fences are fairly common, especially around the larger, wealthier homes. But there are many homesteads with none.

An increasing demand for fuelwood coincides with declining resources. A series of interventions have been tried which have improved woody-biomass production. The lessons from the successes are important. Among these are:

1. Chiefs banning the use of over-grazed or eroded land, leading to rapid improvements in pasture and woody-biomass regeneration.
2. The experience of promoting black wattle for tannin production demonstrates a willingness to plant and tend trees, even on rangeland, if there were clear benefits, and the flexibility of communal land-use for small-scale individual use of trees.

3. Leaving "grass strips" between fields stopped erosion on arable lands. This met with success because it had the support of the traditional authority.
4. Grazing-demonstration areas were introduced in 1984. With the agreement of a chief, an over-grazed and eroded area is fenced off and a limit is set for the number of cattle to be grazed. Recovery of the grass has been quite dramatic, accompanied by improved quality of cattle and woody-biomass regeneration.

These successes demonstrate first, the importance of gaining the support of the grass-roots authorities for successful innovation; second, that popular attitudes and responses to trees are adaptable, according to the perceived economic benefits; and finally, that technical and environmental problems are by no means insurmountable.

Policy to improve fuelwood supplies from rangeland must be primarily as an additional benefit to a programme that addresses other priorities. These must be the contribution of trees towards ensuring a sustained and improved rangeland (without accelerating differentiation in access and cattle-ownership) and the provision of attractive tree products – poles and various forms of fencing.

The Rangeland Management Officers and Forestry Officers are already adopting the strategy of using existing power structures and proven technology. The first is evident in the demonstration areas; the success of these will benefit the local chief more than anyone else, in the short-term. The second is to be found in the promotion of wattle-management in the Middleveld and Highveld. Present in most areas, wattle spreads and grows rapidly, provides fairly good poles and fuelwood, and conserves soil.

These two strategies are the basis for combining rangeland management with fuelwood and other tree products where the natural woodland has been cleared. We will examine each in more detail, looking at the constraints and potential that exist.

Grazing demonstration areas (GDAs)

These should not only be seen as a means to restore the range and improve herd quality. The rapid re-growth of woody biomass on land which is cleared and then grazed with limited stock, is clearly visible on the GDAs. This poses both problems and opportunities.

The problems are that the bush may take over, suppressing grass growth and preventing access to cattle. The potential, however, is enormous. Here you generally have rapid annual growth in woody biomass together with the prevention of soil erosion. If the bush encroachment can be contained, it will not suppress the growth of grass. If managed, it can provide fuelwood and unlimited thorny-brush fencing, so providing a partial answer to tree-planting around the house, in-field browsing and lack of fencing. With the annual thinning of bush on improved grazing areas, a supply of fuelwood and fencing can be sustained at the same time as opening up possibilities for the gradual re-establishment of a limited amount of canopy trees in the rangeland. This would involve selective thinning and possibly some enrichment-planting. The latter would not involve the same problems of survival because lower numbers of livestock would prevent browsing. It may also be possible to introduce tree species that provide fodder.

To make these initiatives work properly would require a sharing of knowledge, and joint planning and operations between forestry and rangeland management. There would also have to be a way of demonstrating that trees do not necessarily harm grass and may indeed accelerate its growth. Furthermore there would have to be the organization of sufficient labour and the management of thinning and extraction. Chiefs would have to play a central role but the products available could constitute part of the returns for labour. If unemployment rises, migrant labour declines and more men live on SNL, it will be interesting to see whether "rangeland management" can be constructed as a new "male" task to complement herding (which with more fencing, will impose fewer labour demands).

The policy direction outlined above would begin to combine woody-biomass and rangeland management with fuelwood being just one of the benefits.

Wattle Management

Wattle occurs in a variety of locations and under a variety of land-tenure and management systems, in the Highveld and Middleveld, with many patches being self-seeded. The Forestry Section focus on wattle is therefore reasonable, stressing in particular its management for tannin and charcoal production. Opportunities for earning cash are seen as the best route to involving people in woody-biomass management.

Wattle has the advantage of not requiring changes in burning practices, but is not a suitable species for providing general tree cover because of its tendency to spread. The best use of wattle is in concentrated areas or to serve specific purposes; to complement the wood products and rangeland improvement envisaged as arising from the GDAs. The management of wattle should concentrate on areas of wood shortages or where the establishment of tree cover is required for soil conservation, banks and wind-breaks.

The critical problems are labour and management. Some of the wattle lots are in a poor condition because of lack of labour, "incorrect" but rational practices such as the extraction of all thin saplings for fencing, and an inability to restrict access, so that cropping is exceeding yield. In areas of severe wood shortages, the orders of the chief are not always respected.

The management problems could be eased by increased forestry training and more forestry diplomats to advise chiefs and Rural Development Councils. The problem of labour could perhaps be approached by charging labour-time for the collection of poles and fuelwood. This would then leave the problems of policing and fencing, particularly against traders. The potential for grazing among wattle depends on fairly intensive management, because of its tendency to form dense thickets.

While labour and management form the core of an "on-range" strategy, their success depends upon the increased provision of wood from non-range land. This would certainly involve the removal of some land from communal access and its fencing. Problems of management, acquiring large tracts of land and large amounts of fencing, mean that the Forestry Department is encouraging small, fenced woodlots to serve a specified number of homesteads or individual homesteads. Without fencing, both have proved complete failures because of browsing. With fencing, it seems an important policy, devolving much of the control over the woodlot on to those who will benefit.

Alternatives have to be found in order to meet a growing urban fuelwood energy demand which is impinging upon the rural resource.

DEDZA-NTCHEU HIGHLANDS, MALAWI

The Highlands are "rugged", containing much bare rock. Soil

erosion is a serious problem resulting from steep slopes, high water run-off and low vegetation cover as a consequence of extensive cultivation. The area is largely deforested. Farming opportunities are limited, given the steep slopes, cool climate, eroded soils and restricted resource base. The already high population density has been further exacerbated by a growing influx of refugees from Mozambique, escaping from the South-African backed terrorist campaign.

The restriction of marginal land for cultivation and deforestation makes the search for fuelwood a serious problem, particularly for the women upon whom this burden falls. Scavenging for twigs and residual wood provides the safety-net for domestic energy needs. However, there are two important woody-biomass resources situated in this area – the Chongoni and Dedza Mountain Forest Reserves. With a worsening fuelwood situation, people are travelling over 60 kilometres by scotchcart to purchase wood that has come from the forest. Here imaginative innovations in forest and natural woodland management can be found to help meet the serious fuelwood problems of the Highlands.

The forest was established to serve the timber industry and there is a large mill nearby as well as many local pit-sawyers. Mindful of the fuelwood shortages in the area, the Forest Officer has made serious efforts to show the local population that the Forest Reserve is not just state land from which they have been alienated but that they can benefit from it. Foresters cut and transport wood to selling points around a widening hinterland surrounding the forest. Selling points have also been established within the reserves. Illegal cutting still occurs, another measure of the fuelwood shortage, but changes in management practice have reduced the level of uncontrolled and illegal cutting.

The most important innovation to meet the growing demand is the adjustment in cutting practices. This involves a policy of combined pruning and clear felling around the whole periphery of the forest rather than at specific sites. This gives people easier access to the off-cuts. This practice was introduced in 1986 and has a mutual benefit to both parties: people get access to the reserve and the Department of Forestry is able to reduce the potential damage such a high demand would create on forest stocks.

The initial aim of the forestry project in this area was to produce timber self-sufficiency given a projected national shortage. A planting target of 400 hectares per year was established with the aim of supplying the large sawmill nearby. At the beginning of the 1980s budgetary difficulties led to a reduction in project resources. In part as a consequence of this, the biggest forest fire in the country's history broke out in October 1984, affecting 1,500 hectares. Although this was a setback to the aim of self-sufficiency, it provided an important resource of fuelwood to meet increased demand. By 1986, about 800 hectares of the partially burnt timber had been cut and sold. By the end of the decade, this particular resource will have disappeared.

Aware of the problems that the loss of this resource and the increasing fuelwood demand raise for the future, planting practices have also changed. Instead of being planted in blocks, trees are now placed around the periphery to build up the buffer zone. In addition, instead of clearing the indigenous wood to plant pure stands of exotic trees, certain areas of these are now being interplanted with the indigenous trees so that the faster-growing exotics will be preferentially harvested for fuelwood. This whole policy is providing vital protection for the country's important indigenous forest resource. Management of indigenous woodland is also being improved and pruning indigenous trees creates a stronger main stem as well as providing fuelwood offcuts.

Nurseries in the forest reserve are selling seedlings to farmers and are encouraging tree-planting. To date however, the species available simply reflect those trees grown within the reserve. Fruit trees, trees for fodder, and other species which meet the particular needs of the surrounding farming populations are required if the forest reserve is to help tackle the wider problem of diminishing woody biomass.

Having explored a number of case studies from within the region, we will try in the following chapter to assess what all this means for developing a new approach to the fuelwood problem.

REFERENCES

1. See A. Millington *et. al.*, *Biomass Assessment* (London: Earthscan Publications, 1988).
2. W. H. A. Tod, *A Comparison of Smallholder Agricultural Development in Kenya and Malawi*, Occasional Paper No. 7 (Edinburgh: Centre of African Studies, University of Edinburgh, 1984), p. 42.

3. O.A.H. Msukwa, *Agriculture and nutrition*, a paper presented at the conference "Malawi: an alternative pattern of development" (Edinburgh: Centre of African Studies, University of Edinburgh, 1984).

4. Government of Malawi, *National Sample Survey of Agriculture* (Lilongwe: Government of Malawi, 1970 and 1984).

5. Energy Studies Unit, *Malawi, Flue-Cured Tobacco Energy-Use Survey* (Lilongwe: Department of Forestry, 1986).

6. Government of Malawi, *Tobacco Sector Study* (Lilongwe: Ministry of Agriculture, 1983), p. 2.

7. Energy Studies Unit, op. cit., 1986.

8. Government of Malawi, op. cit., 1983.

9. Energy Studies Unit, op. cit., 1986.

10. D. Hancock and G.K. Hancock, *Cooking Patterns and Domestic Fuel-Use in Masvingo Province. An Analysis and Possible Options for Decreasing Fuel Consumption* (German Appropriate Technology Exchange, 1985).

11. See M. Froude, "Veld management in the Victoria Province tribal areas", *Rhodesian Agricultural Journal*, Vol. LXXI, No. 2, 1974; and J.P. Danckwerts, *A Socio-Economic Study of Veld Management in the Tribal Areas of the Victoria Province* (Salisbury: Tribal Areas of Rhodesia Research Foundation, n.d.).

12. Ecosystems, *Shinyanga Rural Integrated Development Programme*, Vol. I, 1982.

5. The Secret of Success

ADDRESSING THE CAUSES

In Chapter 1 we examined some of the problems associated with existing attempts to solve the fuelwood problem. This highlighted the limitations of technical innovations such as improved stoves, peri-urban plantations, improved residue use for domestic energy consumption and new and renewable technologies. Subsequently we outlined a different approach and then went on to explore how something similar was already being developed on the ground within the region. What lessons can be drawn from these experiences of improving woody-biomass management within the SADCC region? The first and most obvious point is that it is no good identifying one small symptom, such as a fuelwood shortage, when the causes of the problem are more deep-seated and far-reaching. The difficulty in gaining access to fuelwood is one small manifestation of a broader crisis of rural development and, in many cases, of a declining subsistence base in a period of rapid transition. The key to addressing the causes, rather than the symptoms, is to examine ways of improving land-use management. SADCC has stated as one of its key priorities, the "introduction of land utilization and management practices which are geared to maintaining and improving the productivity of cropland, grazing land and forests".[1] This volume suggests some ways in which this can be achieved.

Improving land-use management involves first identifying and evaluating the land-use and woody-biomass management systems in the specific location. From there, an analysis of the constraints and opportunities for improving the land-use management, and in particular the woody biomass component, can be developed.

In the medium and high potential zones such as the Lilongwe Plains of Malawi, or just outside the SADCC region in the highlands of Kenya, a high population density on good land has led to the growth of woodlots on individual farms for marketed pole production. Indeed, a recent World Bank study has argued that higher population densities are definitely associated with a greater proportion of farm land being managed woody biomass.[2] Detailed survey evidence from Kenya tends to confirm this.[3] However, this does not imply that the response is adequate to meet the high demands placed upon the woody-biomass resource. The key forms of external support required are the expansion of existing agroforestry systems and the facilitation, if and when required, of the removal of any bottlenecks along the chain from supplier to consumer. The constraint lies in the limited and decreasing amount of land available to many of the poorer farmers. A general willingness to improve the agroforestry component on the farms requires the more widespread availability of those tree species which meet the farmers' primary requirements.

In the drier areas of the Masvingo Province in Zimbabwe, we have seen two initiatives aimed at radically transforming land-use management practices by consolidating and demarcating areas for settlement, and both arable and pastoral production. In the case of Mwenezi this involved supporting a local initiative, while with the CARD programme in Gutu this was initiated from above but aimed at involving the local population.

There is certainly much less known about ways to improve woody-biomass management in the drier regions. However, the Shinyanga and Swazi Nation Land examples provide us with some pointers for success. In both areas there have been interesting community initiatives to encourage natural tree regeneration by limiting or excluding grazing for a period of time until the regrowth is sufficiently well established. In Swaziland also, the encouragement of commercialized tree growth, in this case of wattle, has also proved successful.

One of the key problems identified by many people is the difficulty of improving land-use management under communal tenure. Communal lands frequently confront problems of increasing intensity of use combined with a decline in the effectiveness of management systems. Increased pressure makes grazing a priority, often at the expense of trees. On communal land, the challenge is to replenish the supply of fuelwood by demonstrating the contribution that trees can make to other land-uses, particularly to aid the feeding and watering of livestock. Evidence from within the region shows that with some control on livestock numbers and movement (ranging from total exclusion from a given area to rotational grazing or specifying livestock numbers), woody-biomass regeneration can be rapid and sustained. This suggests that provision of fuelwood from a land-use strategy consisting of *cattle among trees* is both feasible and imperative.

Improving the management of natural woodland depends on a better understanding of the primary uses of trees by rural communities, be it on communal land or where the state takes over the indigenous woodland and turns it into a gazetted forest area. In the latter case, a variety of management techniques can be explored. Rotational periods of harvesting can be varied depending upon the type of forest products in demand, be it large or small construction poles, charcoal or fuelwood. The experience of the Dedza–Ntcheu Highlands forest reserve was instructive here.

Improving the management of natural woodlands means encouraging regeneration, in particular by providing protection from damage by fire or livestock. Early burning during the dry season can reduce fire damage and in some areas fire can be altogether excluded. Controlled grazing or the complete exclusion of cattle from certain areas are further measures to be considered depending upon the specific ends in view.

CONTRASTING APPROACHES

Let us now try to derive some general conclusions from the approach that we have been describing, and which are highlighted by the regional case studies. In order to make the new approach stand out, we have compared it with the traditional institutional fuelwood approach in Table 5.1. As can be seen, the result is a series of dichotomies. We can usefully examine some of the project characteristics in greater detail.

Indirect Approach

The successful cases of intervention seem to be those which tackle the problem in an indirect way. The concept of the indirect approach has several dimensions. Its basis is not simply tackling the symptoms of the problem, in this case the manifestation of a fuelwood shortage. Rather, the way to get at the fundamental causes is to discuss with the local community their own pressing concerns. This will reveal not

Table 5.1 Contrasting Approaches to Fuelwood Interventions

PROJECT STAGES/ DESIGN FEATURES	PROJECT CHARACTERISTICS	
	New Approach To Woody-Biomass Management	*Traditional Institutional Approach*
Intervention	Indirect	Direct
Capital cost	Low	High
Technology/ Management	Multiple purpose (e.g. integrated land use)	Single purpose (e.g. tree plantation)
Output(s)	Multiple products	Single product
Activity	Immeasurable	Measurable
Environmental impact	Positive	(Negative?)
Niche	Integrated, diffuse	Firm project boundary
Objectives	Broad goals	Narrow goals
Group Focus	Female and Male	Male

only their prioritization of the problems but will also ensure their involvement from the beginning, in strategies to find solutions to the problems that they define, and which are closest to their hearts. This is an essential feature of a genuinely democratic development strategy. Thus, by assuming that the problem is one of improving land-use management rather than of growing more fuelwood trees, one achieves an understanding of the local participants' own priorities and establishes the foundations for an integrated approach. As a report for the 1988 annual SADCC conference expressed it,

> Most of the problems are inter-related and impinge on all sectors. Often they are due to the management systems which are geared only to the maximum production and not sustainable productivity.[4]

A second important dimension of the indirect approach is that government agencies should encourage people to do more themselves, in the sense that the people rather than the government take on the responsibility for bioenergy production. Here, local community groups and non-governmental organizations (NGOs) have a vital role to play. They can help tap the resource of local indigenous knowledge and bring it into the development initiative. Although indigenous knowledge may exist in pockets, it is not always widespread within the community.

An important illustrative case is provided by one NGO, ENDA-Zimbabwe, in Masvihwa. Prior to launching a community-based tree management and extension project, the existing indigenous ecological knowledge was systematically recorded. This provided an inventory of local tree names with their scientific equivalents explaining their various end-uses, reproductive cycle, local growth conditions and the quality of the wood. In addition, information was collected on people's rights to trees and upon the effects of deforestation. The format chosen for this people-based intervention was a series of open seminars within the community, which helped determine the eventual choice of species to be grown.

The project involves both tree-extension activities and woodland management. These enable local community political structures to devise woody-biomass management plans. The programme will be founded upon the existing knowledge base, with certain community workers receiving short training courses to improve their skills. These workers will convey their new found skills to the local farmers. The evidence thus far is that people appreciate this external impact which helps to consolidate and generalize the knowledge base.[5]

The case for producing fuelwood indirectly by meeting people's higher priority needs from trees has been consistently argued throughout this volume. A corollary to this is that a woody-biomass component can be integrated into non-tree specific projects. Some work has been done on this by the ETC Foundation in Kenya.[6] An attempt has been made to demonstrate that integrating a tree component into a series of development projects can be beneficial to the already existing project goals.

In the case of a dairy development programme, for example, there was difficulty in ensuring the supply of crude protein in the animals roughage. If sufficient crude protein is not available, the feeding system is deregulated and milk production is adversely affected. Introducing fodder trees to the scheme could help tackle this problem. *Leucaena, calliandra* and *sesbania* species have a high percentage of crude protein in their leaves and they will grow in the areas where the development project is situated. Using fodder trees in zero grazing systems is already being tried in an agroforestry project in Lushold District in Tanzania.

Low Cost

The indirect approach can frequently be a low-cost option. In 1982, the Rural Afforestation Project was launched in Zimbabwe. Initially, the emphasis was on growing seedlings in government nurseries for later sale to the population. However, the project discovered that the cost of seedling production was 1,000 per cent higher in government-run nurseries than in decentralized school and community nurseries. The emphasis switched accordingly. Encouraging people to grow more trees themselves is not simply a more cost-effective option, it is also one which is likely to ensure a wider level of impact. Other low-cost ways of encouraging more tree planting include the use of cuttings which are particularly useful for establishing hedges and the use of "wildings" or wild seedlings.

There are many dimensions to this element of low-cost intervention. Agroforestry combinations which include nitrogen-fixing trees such as *acacia spp* can increase crop yields and reduce the cost of chemical fertilizers. Perennials have several advantages over nitrogen-fixing annuals such as alfalfa and clover. They are often active throughout the year which increases total productivity. Their root systems absorb nutrients from deeper in the earth, reducing competition with crops. They offer protection against erosion and provide other benefits to the farmer. Trees which provide fodder release land for uses other than the supply of fuel.

Generally, interventions should aim to have low unit-costs for the government concerned, low recurrent costs, and a low reliance upon imported items. This is vital given the economic difficulties facing many African states. In this way scarce resources can be spread more widely and used to best effect. Thus, the impact will be appreciably greater.

Multi-purpose management

The traditional forestry approach was concerned with single-purpose tree management in plantations or protected, gazetted forest reserves. The new approach is directed more towards multi-purpose management based upon integrated land-use. Intercropping of trees with crops is one such example and with the right combination it is possible to achieve higher production than if the crops and trees were grown separately. The International Institute for Tropical Agriculture has been developing new alley-cropping techniques better adapted to agricultural mechanization. These developments obviously will have institutional implications for governmental initiatives and support services. Does agroforestry belong to Agriculture or Forestry, or (the ideal solution) to both?

Multiple products

Instead of growing a product with a single end-use in mind, such as fuelwood plantations, tree species can be chosen which meet multiple end-uses. Table 5.2 provides a list of the many products that can be derived from a number of trees adapted to the more hostile environment of arid and semi-arid lands.

Other features

One potential problem with the indirect approach is that its success may be hard to measure using conventional methods of project appraisal. An important reason for encouraging people to grow more trees themselves is to try to improve the sustainability of subsistence production. As yet, it remains difficult to assess the extent to which this has been achieved. Generally, the aim is to create a positive environmental benefit. This is likely because the approach is based upon integrated land-use, a concept which is rather different to the firm project boundaries of the traditional approach. Finally, the goals tend to be broad rather than narrow and with a group focus on both women and men, rather than being exclusively male-orientated.

In many parts of the SADCC region there are areas of fuelwood shortage for people living on communal tenure land adjacent to the areas of fuelwood surplus. As we have noted in the case of Gutu in Zimbabwe, there are already examples of resource-sharing taking place between those people living on communal tenure land with a fuelwood shortage and nearby areas under different ownership and land-tenure systems. Therefore, what options exist here for government intervention to help rationalize this process and ensure that it is planned in a manner that does not lead to environmental degradation and conflict? The answer would appear to offer an important channel for tackling the fuelwood problem in certain

Other Remarks (diagonal column headers, left to right): Combines well with agric. crops · Combines well with · Yields gum arabic. · Widely used for livestock pens. Termite resistant. · Fast growing. · Sericulture/shellac. · Very toxic to pigs. · Subsistence food. · Presscake makes good stock feed. · High biomass yield. · Combines very well with agric. species.

Group	Use	Acacia albida	Acacia senegal	Acacia tortilis	Albizia lebbeck	Cajanus cajan	Cassia siamea	Cordessoria sciulis	Pithecellobium dulce	Prosopis chilensis	Prosopis cineraria
Miscellaneous	Ornamental	•		•	•	•			•		
	Cultural, Ritual, Social										
	Medicinal/Drugs			•							
	Fibers			••							
	Gums			••		•	•			•	
	Tannins (T) Dyes (D)	(T) •							(D) •	(T) •	
	Essential Oils										
	Waxes										
Services	Mulch										
	Organic Manure	••				•		••			
	Nitrogen Fixing	•••	•	•		•••	•••		•	•	•••
	Dune Fixing		•	••							•••
	Soil Conservation	••	••	••	••	••	••	•		••	••
	Live Fence								•••		
	Windbreak								••	•	
	Others										
Wood	Sawn Timber				••		••		•		
	Wood for Utensils & Tools	•		••						•	•
	Building Material	•		••	•				•	••	•
	Charcoal		••	•••	•						•••
	Fuelwood	•	•••	•••	•••	•	•••		•	•	•••
Fodder	Bee Forage			•••	•••				•••	•	•
	Shoots			•	•	•	•				
	Fruit/Pods/Seeds	•	••	••		••			•••	•	•••
	Leaves	••	••	•••	••	•	•	•	•••		•••
Food	Spices (S)/ Condiments (C)										
	Starch							••			
	Oils/Fats							••	•		
	Vegetable					••					
	Nuts/Seeds		•			••			•••		•
	Fruits/Pods									•	

Fair •   Good ••   Excellent •••

Table 5.2 Some leguminous and other nitrogen-fixing woody perennials with agroforestry potential in arid and semi-arid areas.

areas. It presents a challenge but also a great opportunity for better land-use management. Both parties to the resource-sharing process can stand to benefit. The ownership of a tree and the ownership of its yield are not necessarily identical, hence yield-sharing can protect the resource and put to good use what might otherwise be wasted.

One important possibility for resource-sharing between small-scale farmers facing problems of land shortage next to state or commercial forestry plantations, is the "taungy" system. Here, foodcrops are grown within newly established plantations. This helps to reduce the costs of establishing the plantation as the farmers growing crops between the trees reduce the need for weeding and the result is an additional product.

Table 5.3 presents the range of fuelwood resource-sharing schemes within the SADCC region, outlining existing practices and possible options. Resource-sharing offers the possibility for improved land-use management practices, increasing both land values and potential cash income to the benefit of the commercial and communal farmer alike. Quite clearly, this approach is applicable to a wide range of resources including grazing, wildlife and water.

Our argument has emphasized that we must consider woody-biomass management within the agricultural system and not simply focus on tree production. This has important implications for extension work since it implies that efforts should be re-focused away from those aimed at growing individual products, such as fuelwood and towards those that adopt a systems approach to the wood problem.

Table 5.3 Range of Fuelwood Resource-Sharing in the SADCC region: existing practices and possible options

1. Illegal (existing practices)

    (a) poaching with strict legislation and strict policing;
    (b) poaching with strict legislation and lax policing;
    (c) poaching with official legislation but "permission" from the authorities, either as a result of implicit policy or insufficient resources to enforce the law (by government, parastatal, private or customary authorities).

2. Legal (existing practices)

    (a) the granting of free access to woody-biomass resources;
    (b) the granting of access by licence or permit, specifying conditions of gathering (e.g. quantities, species);

(c) the granting of access with payment of a nominal fee (or unofficial payment);

(d) the granting of access by the payment of a more than nominal fee which is less than the replacement cost of the resource.

3. Reciprocal benefit arrangements (existing and possible)

   (a) access is granted on the basis of "good neighbourliness"; for example, to avert absolute fuel shortage and other poverty or hardship;

   (b) tax concessions or other forms of payment for owners of un- or under-utilized resources (e.g. large commercial farmers, the Forest Department) if they enter sharing arrangements with neighbours on poorer land;

   (c) rents paid by resource-poor farmers for the managed use of un- or under-utilized wood resources (e.g. commercial farmland, state forest or rangeland);

   (d) leasing of state land on the condition that it is managed productively;

   (e) use of state land by smallholders with sharing of costs and benefits between them and the state (e.g. joint-venture tree planting);

   (f) land reform and redistribution to favour people facing resource scarcities, with compensation for the dispossessed.

There are also methods of resource management that fall under this heading, notably:

   (g) planting of high-yield buffer zones around protected forest reserves to deflect destructive, illegal attacks on them.

SPREADING THE WORD

Foresters can no longer remain hidden in the forests. This is the message that clearly springs from the new approach. Forest management will naturally remain an important task, but it can no longer be an exclusive one. For too long foresters have practised their skills in isolation from the local community, constrained by a job definition limited to conserving and maintaining the output of the forests. The tree needs of the majority of the people have been neglected as a result. Studies in Zimbabwe, for example, have shown that uncleared natural vegetation on the large-scale farms (63 per cent) and communal farming areas (23 per cent) account for 86 per cent of the country's wood requirement.[8] People meet their needs mainly from outside the gazetted forests.

Simply coming out of the forests is not enough, however, as experience has so far shown. Not all of the peri-urban mini-plantation schemes and village woodlots first developed, with their high costs and a single-purpose product, have been a success, and few had the cash to purchase the poles they produced. No efforts were made to discover people's real tree needs.

The main goals of forestry extension must be to encourage people to grow more trees themselves, and to improve the management of existing tree resources. An important starting point is understanding the multiple roles that trees play within the environment in which people live and work. Extension workers will need to be trained to work within the landscape, to recognize and identify the evolving production systems. Bottlenecks within the woody-biomass component of these systems can then be identified and overcome. Appendix I looks in some detail at specific agro-forestry options that can be employed to encourage improved woody-biomass management.

Extension workers will need training in how best to use the farmer's knowledge, technical skills and practices. Changing the emphasis from seedling distribution to seed distribution is a concrete example of how this can work.[9] The distribution of seedlings from centralized nurseries is not terribly effective over long distances, and is also expensive. Farmers tend to distribute their own tree seeds to relatives and friends in the local community. Therefore, when new species are introduced, greater emphasis should be given to distributing the seeds, rather than seedlings, to large numbers of farmers and the "neighbourhood effect" described above would spread them further still. Instead of growing the seedlings in expensive government-run nurseries, more advice could be given to farmers about how to improve their own farm nurseries through manuring, watering, weeding and root pruning which all help to ensure that a higher number of seedlings survive.

Extension workers can encourage farmers to experiment with new species themselves. The farmer will not want to invest a lot of his land and labour in a major planting programme until its effectiveness has been proved. He or she will first want to see how a few such trees prosper in the local farm environment. Hedges are often a good initial site to choose as this does not entail giving up existing arable or pastoral land. New species will inevitably be subjected to the farmer's critical eye.

Cash crops and subsistence food crops have a clear production cycle, unlike woody biomass. The extension agent can help identify optimum rotation periods for woody biomass on the farms. Careful attention needs to be paid towards establishing the best rotation periods given the specific wood needs of the farmers. Fuelsticks and fodder can be provided on an annual rotation, poles and fuelwood between two and four years, while timber production will take somewhat longer. The choice of species will have to be made according to the uses required as well as the rotational periods. It may often be best to begin with those woody-biomass management techniques that bring quick rewards as this will maintain the farmer's interest and enthusiasm.

For extension workers to be effective in this new approach, they will need a thorough understanding of the role of trees within land-use systems. There are generally three main woody-biomass systems on the farm: trees on crop or pasture land, hedges, and woodlots, in high population density farming areas. Looking first at trees in cropland, extension workers can increase production by improving the pollarding system. Pollarding at the start of the rains will reduce shade to the crops and ensure the best use of the rain. Agroforestry cropping systems can be developed. In the hedges, depending upon the needs involved, fodder or fuelwood can be produced and the fastest growing species can be recommended. In the woodlots, the extension officer can help advise the farmer on the best techniques to produce the type of commodity suited to the market demand.

Generally, the tree species available from government agencies throughout the region reflect the plantation bias of the forestry profession to date. Seeds and seedlings of species able to meet the farmer's specific requirements are a high priority. Developing tree species useful to the farm and ensuring their widespread dissemination is a necessity. These farm trees would be perennial short-rotation crops providing soil enrichment, fodder and fuelsticks. Depending upon the actual farm context, they could be grown in woodlots, boundary hedges or on cropland. New farm tree species to fit this bill include *calliandra calothyrsus, mimosa scabrella, leucaena leucocephala* and *gliricidia sepium*. For this new approach to work properly, however, extension agents will have to move away from a situation in which targets are set for the number of trees to be planted rather than the number of farmers to be reached.

A far humbler approach on the part of extension workers will be required for them to understand and appreciate local indigenous knowledge and practices. For this to occur they will have to learn to listen, a most difficult task. In particular women have an important role to play. As Victoria Chitepo, Zimbabwe's Minister of Natural Resources and Tourism, has commented, "International agencies and governments have everywhere ignored the vital part that women play in caring for the environment. Their voice, like their knowledge and experience, is simply not heard."[10] There is no formally established body of knowledge concerning woody-biomass management practices. Developing that body of knowledge is a crucial challenge. Willem Beets had stated the problem in stark terms:

> Systematic research for generating suitable technologies incorporating woody perennials in arable land use practices is still in its infancy. This means that, although many possibilities can be identified, there is as yet little solid and proven technology nor data on production. More research is urgently required in all disciplines related to agroforestry, not only in plant sciences, agronomy and farm management but also in socio-economic and institutional investigations.[11]

Systematic work is now being undertaken, notably under the auspices of the International Council for Research in Agroforestry (ICRAF). They have been developing a methodology of agroforestry diagnosis and design in order to examine land-use management problems and consider particular agroforestry solutions. This involves a series of stages: prediagnostic, diagnostic, evaluation, redesign, planning and implementation in an iterative process. In Zambia, experiments have taken place, using this procedure to build upon the indigenous technical knowledge contained in the *chitemene* system of shifting cultivation.[12] Under the modifications made within the cropping system, a ten-fold increase was recorded in the carrying capacity of the land.

Research efforts are therefore urgently required to back up extension work. In a number of SADCC countries, attempts are being made to catalogue tree species, their allocations to climatic zones and their potential uses in agroforestry. B.K. Kaale has produced one such important study for Tanzania.[13] At an SADCC Regional Planning Workshop on Forestry Extension Training held in Malawi in June 1987, one of the high priorities to emerge from the meeting was that:

> Research has to be reorientated to meet the needs of this new approach. In particular there is a need for research on indigenous woodlands and species, on indigenous knowledge and practices and on the wider cultural, social and economic issues of the role of trees in production systems.[14]

The need for an active extension service is too great for it to await the completion of all the necessary research and development at a scientific level. Extension workers have to begin to mobilize local farmers immediately, in order to accumulate the evidence that only emerges from a careful consideration of local experience.

Mobilization means the creation of new opportunities for rural people. Successfully mobilized, the local farmers must acquire an understanding of their own situation, its potential for change and their own possible role in that changing process. It requires that their implicit knowledge of the woody-biomass management system be made explicit, that the opportunities and constraints of that woody-biomass management system be clearly established, and that they are encouraged to organize new management strategies of woody biomass within the context of farm production.

No matter how skilful the extension workers are, and how sensitive they are to local attitudes and concerns, no project can work in the long run if it does not engage the full enthusiasm and commitment of the people whom it is supposed to benefit.

Involving local people at a project's inception is undoubtedly one of the keys to effective action. Rather than convincing villagers that they have got a fuelwood problem, and persuading them that growing trees or rangeland management would be a viable solution, it is far better if they themselves are left to define the problem – and come up with what they think is an acceptable solution. Usually, this will be far more in tune with local conditions, and much more likely to succeed as a result.

Instead of being the main driving force behind projects, the extension agency can thus become a catalyst and a facilitator. This will not be an easier role to play; if anything, it will be more difficult as it requires a lot more patience and often a completely different attitude on the part of the extension worker.

The Mwenezi land-use management scheme in Zimbabwe provides an extreme example of how this approach can work. Here, the initiative for the project came from local village leaders rather than from outside. It was they who identified the problem of land degradation, and they who decided that separating arable and pastoral areas, with a system of rotational grazing, would be the best way of dealing with it. Only then did they approach the government for assistance in providing technical inputs and advice.

Obviously, waiting for communities to come forward with requests for assistance cannot form the basis of a national programme for communal biomass management. But it does demonstrate a different kind of approach to the whole problem, a distinct contrast to the standard one of diagnosing problems from outside and designing programmes to disseminate "solutions".

What is needed, perhaps, is a blend of the two. Government agencies clearly have a role to play in stimulating local action. The point is that they have to know how far this role should go. Wherever possible, they should take a minimalist approach, doing what is needed to initiate local discussion and providing the inputs necessary to get projects started, but leaving as much as possible to the communities themselves. A recent FAO forestry report suggested:

> Small tree growing programmes may be more successful when forest services offer technical support while NGOs implement the activities with the residents.[15]

In Zimbabwe, this approach is beginning to be formalized under a programmed developed by the Department of Wildlife. It is called the Communal Areas Management Programme For Indigenous Resources (Project Campfire), and it is intended as a means of involving communities in deciding how best to manage their local resources. It seeks to place custody of resources with the local communities and provide a framework within which they may manage and benefit from these resources.

Obviously, this kind of approach is not guaranteed to work. Projects will still depend on getting the technical package right, resolving the conflicts of interest that inevitably occur within communities, and having the flexibility to learn from any mistakes that are made. They will also require a considerable amount of patience. But it does represent an important new line of attack, and one which holds considerable promise for the future.

Table 5.4 indicates the broad range of extension opportunity. The evidence would suggest targeting woody-biomass management extension towards the functional group and farmers' organization approach rather than the more traditional scheme, commodity or technical change approach.

To date, extension initiatives in the pastoral areas have at best emphasized technical change, although the worst initiatives, increasingly abandoned by both government and donors, have been focused on the scheme initiatives. The success of woody-biomass management on individual farms must be the starting point for future extension activities, a starting point that attempts to push the

Table 5.4 Approaches to Woody-Biomass Management Extension

Approach	Unit primarily benefiting from extension efforts	Methods used by extension for influencing behaviour
Scheme	State/plantation owner/ urban population/possibly the tenants	Instructions for compliance with strict rules and regulations
Commodity	State boards and participating farmers	Commercial, technical and administrative instructions within monopolized market and/or compulsory wood-production systems
Technical Change	Individual farmers	Dissemination of technical information
Target Category	Categories of population chosen for the similarities of their needs or opportunities	Provision of carefully selected information focused on specific needs
Functional Group	Groups formed by joining efforts and resources to achieve a shared goal	Mobilization, organization, technical and resource support training
Farmer's Organization	Permanent organizations with institutionalized development objectives. Independent and self-managed	Cooperation, negotiation, promotion of self-management services

successful initiative to more marginal areas. Frequently, any strategy that seeks to enhance wood production must locate the wood resource close to the settlements of the population. This requires a strategy for either planting by individual peasants or community woodlots.

Extension activities cannot be left to foresters alone. Agricultural extension is far more developed and has a greater capacity than the fledgling forestry extension services. In a number of SADCC countries, forestry extension services are only just beginning to be put into operation. Beyond the obvious limitation of there being too few trained people for the work involved, there are other powerful reasons for widening the institutional base. This is certainly necessary if a more integrated approach to rural development and land-use management is to occur. On many occasions, conflicting advice has been given by different agencies. Sometimes, departments of agriculture have recommended clearing all of the woody biomass in order to develop crop cultivation. Yet there are situations in which these conflicting needs can be harmonized. There has been a traditional view that grazing land and trees compete. Yet to pose the problem as an alternative between grazing land and livestock *versus* trees and tree products is a false dichotomy, since they are interdependent. An unintentional benefit of grazing-land recovery programmes, both in Swaziland and in Tanzania, has been the regrowth of woody biomass which need not hamper future grazing land and can meet a variety of needs for wood in the community.

One of the greatest problems faced in spreading the word about the possibilities for improved woody-biomass management is our ignorance of the cultural and gender issues at a domestic level. In matrilineal societies, for example, male insecurity of land tenure may profoundly effect willingness to engage in tree planting. A central problem in many societies within the region and beyond, is that women have the responsibility for fuelwood collection but are often not permitted, under custom and tradition, to plant trees. Conversely, fuelwood collection is a woman's responsibility so it is of no interest to men, yet it is frequently men who exercise the sole right to plant trees.[16] Trees are generally regarded as men's property.[17] Finding ways out of this dilemma will be a key problem for extension workers. One solution may come from the fast-growing fuelstick and fodder trees, namely a *redefinition* of a "tree" as a "crop", which can sidestep cultural taboos against women planting them. This is one avenue which has been explored in the Kenyan Woodfuel Development Programme (KWDP). Under Lukya society law, in the Kakamega District of Kenya where much of the KWDP work has been carried out, the origins of these taboos are that land ownership disputes were resolved by adjudicating on the basis of male tree-planting on the land, in a social context where possession of capital assets was confined to men.

These questions of land/property ownership are dimensions of domestic power relations which are defined by gender and culture.

In addition, there is the pervasive reality of the absence of men from the rural household for extensive periods of migrant labour. In the case of Zimbabwe, Olivia Muchena has mapped out the implications of male migration for women in the rural household.[18] Women have to take over the responsibility for farming tasks previously undertaken by men. In addition, after Independence, an expansion of educational provision meant that children were no longer available to help the women to the same extent with domestic tasks, including fuelwood collection and agricultural production. The burden upon women increased and the extension services have not been providing the necessary training or credit and other services which could lighten this load and stimulate women's development potential.

A central problem is that in most societies in the region, women have no access to their own land in any circumstances. This is because they are effectively classed as legal minors. A major UNICEF survey of women in Zimbabwe showed that 99 per cent of women wanted the existing land-tenure system to be modified, or abandoned altogether.[19] To sum up, women face the problem of being the daily farm-managers, yet the cultural overlay of male decision-making on the farm continues. These leave the women in a situation of management without authority. Women have hardly any access to the local chief and are therefore unable to refer their problems to the traditional authorities.

It will take more than an improved and more aware forestry-extension service for this situation to change! Clearly, amending the sexual division of labour within the household will take a very long time. Yet properly targeted and organized extension work can contribute towards overcoming some of these barriers. As women bear the burden of fuelwood collection, often spending many hours finding and collecting wood, they will be the potential beneficiaries of enhanced, on-farm woody-biomass production. More women extension workers and a greater awareness of the cultural and gender constraints that effect extension activities will all help.

Spreading the word requires more than an improved and expanded extension service, although this is certainly a priority. Public awareness needs to be raised. Many countries now have national tree-planting days, with a positive lead being taken by the head of state. Constant publicity in the mass media, on the radio and in newspapers in particular, can have a significant impact. Such at least has been the experience in Tanzania, Zimbabwe and Malawi.

Finally, however, the prospects for "spreading the word" are not just dependent upon political will but are greatly constrained by economic circumstances. The worsening economic situation in a number of countries in the region and, on top of this, the massively disruptive effects of destabilization, have limited the capacity of extension workers to be fully effective in the field. In Tanzania the "Forests are Wealth" campaign suffered severely from transport bottlenecks and financial difficulties.[20] To overcome these problems, a more efficient use of resources is required and the only place to start is by tapping the potential of the local farmer.

REFERENCES

1. SADCC, *Food, Agriculture and Natural Resources*, a report presented to SADCC Arusha, 28–29 January 1988, p. 12.
2. World Bank, *Kenya Forestry Subsector Review: Main Report* Vol. I (Washington DC: World Bank, 1987).
3. P.N. Bradley, N. Chavangi and B. van Gelder, "Development research and energy planning in Kenya", *Ambio*, Vol. XIV, No. 4–5, 1985.
4. SADCC, op. cit., 1988, p. 11.
5. K.B. Wilson, *Research on Trees in Masvihwa and Surrounding Areas* (Harare: ENDA-Zimbabwe report, 1987).
6. B. van Gelder, *Agroforestry Components in Existing Netherlands Assisted Projects in Kenya* (Nairobi: ETC Foundation, January 1988).
7. Source: W.C. Beets, *Agroforestry in African Farming Systems* (Washington DC: Energy/Development International USAID, 1985), p. 60.
8. R.H. Hosier, Y. Katerere, D.K. Munasirei, J.C. Nkomo, B.J. Ram and P.B. Robinson, *Zimbabwe: Energy Planning for National Development* (Uppsala: Scandinavian Institute of African Studies, 1986), p. 63.
9. Some of the ideas discussed in this section have been worked out by B. van Gelder and B. Munslow in a draft paper entitled *An Introduction to Developing and Improving Woody Biomass Systems on Farms* (Nairobi: ETC Foundation, 1987).
10. I. Dankelman and J. Davidson, *Women and Environment in the Third World* (London: Earthscan Publications, 1988), p. ix.
11. W.C. Beets, *Agroforestry in African Farming Systems* (Washington DC: Energy/Development International [a USAID commissioned report], 1985), p. 41.
12. G. Leach and R. Mearns, *Bioenergy Issues and Options for Africa* (London: IIED (draft), 1988), pp. 32–6.
13. B.K. Kaale, *Trees for Village Forestry* (Dar-es-Salaam: Forestry Division, Ministry of Lands, Natural Resources and Tourism, 1984).
14. *SADCC Forestry Extension and Training Strategy* (Lilongwe, Malawi: Planning Workshop on Forestry Extension Training, 8–17 June 1987).

15. FAO, *Tree Growing by Rural People*, FAO Forestry Paper 64 (Rome: FAO, 1985), p. 100.

16. P.N. Bradley, N. Chavangi and B. van Gelder, op. cit., p. 235.

17. E. Cecelski, *Energy and Rural Women's Work: Issues for Discussion* (Geneva: conference paper for the preparatory meeting on Energy and Rural Women's Work).

18. O. Muchena, *Women and Work. Women's Participation in the Rural Labour Force in Zimbabwe* (Harare: ILO report (unpublished), 1982).

19. O. Muchena, *Report on the Situation of Women in Zimbabwe* (Harare: UNICEF and the Ministry of Community Development and Women's Affairs, 1982).

20. Institute of Adult Education and Ministry of Natural Resources and Tourism, *Final Report of "Forests are Wealth" Campaign* (Tanzania: Institute of Adult Education and Ministry of Natural Resources and Tourism, 1982).

6. Locating the Problem Areas

THE PROBLEM OF WOODY-BIOMASS SUPPLY ASSESSMENT

Thus far we have looked at the fuelwood trap and explored ways of avoiding it by developing a new approach to the problem and new methods of dealing with it. Now we turn to examine ways of locating the problem areas. A study by the FAO published in 1983 tried to produce a map of relative fuelwood scarcity for all developing countries.[1] Data for the supply estimates was compiled from the natural growth rates of vegetation and the accessible sustainable supply. Consumption estimates were based upon the available data. By comparing the two sets of figures, the extent of fuelwood scarcity was assessed. Twenty-year projections were then made, based on estimates of population growth.

According to the map subsequently produced, the following results were obtained for the nine SADCC countries. Lesotho and Swaziland were categorized as areas of acute shortage. Deficit situations were located in southern Mozambique, Malawi, east, central and northern Tanzania, the east of Zambia and a broad swathe of territory across Angola that stretched diagonally from the north-east to the south-west. Areas of prospective deficit were Zimbabwe and the remainder of Mozambique and Tanzania. A satisfactory situation was proclaimed for the north-west and south-east of Angola, the west of Zambia and all of Botswana with the exception of the south-western third of the country which was categorized as desert and sub-desert areas in a scarcity situation.

While this represents a worthy early effort to examine the issue and while it served to draw attention to the problem, there are fundamental difficulties involved in employing such sweeping categories over such an enormous area. To be fair, the FAO report did emphasize the weakness of the data upon which the assessment was based. The geographic scale is clearly too large to be of operational use. As we have already seen, even within a given district which has been defined as having serious fuelwood problems, the situation can differ markedly over a few kilometres (see, for example, Figure 4.1). The reality is therefore, infinitely more complex, with stress occurring as an intricate pattern or mosaic. Hence sweeping categorizations do not help us to discover a particular problem area. There is a perennial problem with *averaging* figures, as the *range* of those figures is frequently more important than the average itself.

Furthermore, as we have seen, people rely mainly upon trees outside the forest for their various wood requirements, so there is generally a far higher level of woody biomass available than was first thought. It is also apparent that people face a preferential hierarchy of systems in their choice of the woody biomass available. Given increasing scarcity, they can move down the ladder to lower quality firewood, twigs and bushes, and finally down to other biomass.[2] Hence, the early work tended to dramatize the "fuelwood problem" in too sweeping a manner. It did not help to locate the problem areas.

Assessments of biomass supply in Africa have generally been based upon national forestry statistics. Inevitably, different countries demonstrate a high degree of variability in the effectiveness of their forestry departments and this affects their consequent ability to undertake such a task. The relative degree of accuracy in their assessments will reflect to some extent the resources that they have available. There is no uniformity across national boundaries in assessment techniques, and this creates problems in achieving the necessary compatibility and comparability of data. When making regional or continental assessments, the application of standard techniques is a fundamental requirement, as is the need to carry out longitudinal studies.

A further difficulty emerges from within the departmental ethos of forestry itself. Foresters have primarily been concerned with commercial forests rather than sharing a wider concern with the total woody-biomass resource. Inevitably, this is reflected in differing levels of accuracy in the assessment of the various categories of woody biomass. A much broader approach to biomass assessment is therefore required.

A third difficulty concerns the pace of change taking place in Africa. Rapid increases in population growth lead to the transfer of more and more land to agricultural production. This rising rural demand, coupled with the massive rate of urban growth and the subsequent high demand for wood, means that biomass-supply data soon becomes outdated. Mechanisms are therefore required to ensure a regular monitoring of the changes occurring and an updating of the data base.

In addition to forestry statistics, ecological studies have sometimes been used to inform biomass assessments. These certainly show more potential but again demonstrate two levels of difficulty reflecting those found within the forestry statistics. There is no standardized methodology and there is the problem of keeping information up to date.

One answer to the problems posed by existing sources of information on biomass supply is the use of remotely-sensed data. The great advantage with this approach is that it can produce standardized biomass classifications that are both current and accurate. Repetition of the remote-sensing exercise at regular intervals will ensure an appreciation of the changing biomass supply situation which is vital for energy-planning and budgetary purposes. A number of countries have already undertaken such an exercise: Tanzania and Kenya in East Africa, and Botswana and Mozambique in Southern Africa. However, the value of these studies has been limited as the participants did not have access to digitally-processed data which uses a variety of computer-based techniques to produce a higher level of information than simple photographic material. Many of these earlier studies also relied heavily on outdated and sometimes inappropriate image-processing algorithms.

Of course, remotely-sensed data is not enough to produce a satisfactory biomass-resource assessment on its own. Secondary sources such as existing maps and reports on natural resources have to be used to verify the biomass classes identified from the remotely-sensed data. There also needs to be a measure of ground verification and calibration. Different remote sensing techniques can give differing scales of data.

BIOMASS SUPPLY IN THE SADCC REGION: COUNTRY OVERVIEWS

In a recent study that we carried out for the nine member states of the SADCC, a remote-sensing exercise of regional biomass supply was undertaken in collaboration with the Geography Department at the University of Reading.[3] Based upon this earlier work we will provide a brief summary of the woody-biomass supply situation in each country, focusing particular attention upon the potential problem areas.

Angola

This is a huge country with a relatively small population. It contains over one half of the SADCC region's woody-biomass resources and a wide spectrum of biomass categories can be found. However the districts of Bengo, Luanda and Namibe contain dry inland savanna, dry coastal savanna and arid coastal thicket vegetation of low productivity. These districts face the most widespread problems of woody-biomass supply, particularly around their urban areas. The coastal zone contains the most significant urban conglomerations which gives rise to particular problems but ones which are specific to their locations.

Botswana

This is a largely semi-arid environment, suffering a shortage of rainfall, poor quality soils and a high cattle population. Inevitably, it faces a situation of restricted biomass resources. The area worst affected is the relatively densely populated south-eastern corner, whilst the north contains areas of higher potential woody biomass. With the exception of Chobe and Northern Ngamiland, however, all districts of the country face existing or potential supply problems. Inevitably it is the areas with major urban concentrations that face the most acute supply problems.

Lesotho

Unique amongst the SADCC countries, Lesotho has very few trees at all. The primary reasons for this are its high altitude and location in the southernmost part of the continent. However, its labour-reserve economy has produced, to a certain extent, an ecological profile similar to some of the South African bantustans. It is a measure of the country's extensive environmental problems that within the division of portfolios between the SADCC member states, Lesotho has responsibility for soil erosion. The whole country faces a woody-biomass supply problem.

Malawi

This country exhibits a clear spatial division with regard to woody-biomass supply. The central and southern districts face a general supply problem. Indeed, six of the seven problem areas are to be found here in the Lilongwe and Phalombe Plains, the Dedza-Ntcheu Highlands, the Shire Valley, the southern shores of Lakes Malawi and Malombe, and the plantation areas around Thyolo and Malanje. The seventh area is technically in the north but is exceptional in being two islands in Lake Malawi, Chizumulu and Likoma, which have their own specific problems. Otherwise the mainland areas of the north contain substantial woody-biomass resources.

Mozambique

This country has a generally healthy woody-biomass supply situation with a considerable diversity of resources available. However, the provinces of Maputo, Gaza, Manica and Tete exhibit certain signs of stress with former areas of woodland now degraded into savannas. In addition, parts of the northernmost province of Cabo Delgado face woody-biomass supply problems.

Swaziland

The woody-biomass supply situation of Swaziland is closely related to its four distinct regions which are, moving from west to east, the high, middle and low veld, and the Lubombo Hills. In the highveld, the woody-biomass supply is accounted for predominantly by commercial plantations to which general access is severely restricted. In the middleveld, a combination of pressures on land for arable and pastoral production, plus a lower biomass potential, makes this a key problem area. The low veld and the Lubombo Hills do not currently face significant problems.

Tanzania

One of the largest SADCC countries, Tanzania contains a wide variety of woody-biomass resources. Inevitably, there are marked regional disparities in availability and demand. Nine out of the nineteen districts in the country have areas of existing, or potential, woody-biomass supply problems. These are Arusha, Dodoma, Iringa, Kilimanjaro, Mara, Mbeya, Shinyanga, Singida and Tanga Districts.

Zambia

Like Tanzania, Zambia contains a wide variety of woody-biomass resources. Whilst exhibiting a favourable woody-biomass supply situation overall, notably with the extensive Miombo woodland coverage, problem areas lie in the middle of the country in the Copperbelt, Central and Lusaka provinces. Here, the problem is particularly associated with heavy urban demand. Zambia is the first country in the region to approach a 50 per cent urbanization of the population and this places particular stress on the woody-biomass supply.

Zimbabwe

The white-settler past of Zimbabwe created a situation in which areas with the high woody-biomass supply potential were occupied by white farmers with a low population density while the lower potential areas were consigned as African labour reserves. These reserves made a higher demand upon the woody-biomass resource for subsistence purposes and problem areas include Matabeleland North and South, Midlands, Masvingo and Mashonaland Provinces.

FOCUSING IN

The initial aim is to obtain a broad picture of major problem areas at a regional or national level, using an assessment of the woody-biomass resource base and the pattern of demand based on population densities. The next question is how to focus in on particular problem areas. At this point, a disaggregated breakdown of fuelwood supply and demand is required.

The starting point is a situational analysis and this can be arrived at by a variety of techniques. Energy-planning models can be very sophisticated but the choice of such techniques must be weighed against the other demands being made on scarce resources and available data. Sometimes a less formal approach is called for, using a range of maps, demographic data and estimates of energy consumption. Those with practical experience on the ground can enrich the picture further, indeed it is often the people on the spot who alert the national institutions that a particular area has a problem. Anecdotal reports and observations of broad indicators of stress can all have a place. Just because evidence does not appear in neat, quantitative forms does not mean that it is not important.[4] A situational analysis will indicate the broad problem areas, giving some idea of the degree of severity of the problem.

A particularly powerful tool for biomass assessment is remote sensing using satellite imagery. It is important to combine this with a field verification programme and other documentary sources. Such an exercise has been undertaken for the SADCC region as part of this study and we have summarized its broad conclusions.[5] This situational analysis does not tell us where exactly the fuelwood problems are, nor the real nature of the problems involved. Its essential purpose is to identify localities for further investigation.

A next step is the more detailed disaggregated analysis of an identified problem area. At this stage, much more detailed information will be required. The resource base has to be examined in order to reveal patterns of land tenure, biomass-resource management practices and signs of environmental stress. Following this, patterns of population distribution, economic activity, social differentiation and energy demand can be analysed as well as the general structure of woody-biomass resource management and use (including non-fuel use). The two crucial variables to begin with are population density and the degree of woody-biomass availability. Then follows a detailed assessment of the nature and extent of the problem.

The process of moving on from situational analysis to project identification consequently involves a shift of focus from broad potential problem areas to specific problem sites. This entails looking at the fuelwood system in greater detail over a smaller area, at each stage. The closer one wants to get to the specific form of fuelwood problems that exist in an identified community, the more one needs to know about how fuelwood relates to other aspects of life in that community. Although interventions must be driven out of this relationship, it is necessary to identify what problems exist in order to reach this stage. The process outlined here integrates national situational planning with local project development. This integration is an essential aspect of planning for fuelwood interventions if these interventions are to reach people in a form which reflects the problems they face.

At site level, a simple set of indicators could be developed which experience has shown to be symptomatic of increasing pressure on fuel resources. Clearly, such indicators would vary from country to country. An example of this would be as follows, with the indicators arranged in increasing order of seriousness:

- Women walk longer distances to collect fuel.
- Some crop residues are used in addition to wood.
- Men assist in fuel collection.
- Carts and vehicles are used for fuel collection.
- Dung becomes a major fuel.
- Firewood becomes a commodity that is regularly purchased for home use.
- Fewer meals, or different foods, are cooked.

Care must be taken, however, in the construction of such a list as in certain limited areas such practices may be traditional. For example, in the Kagera region of Tanzania, it has always been the responsibility of the men to provide the firewood. The advantage of such a set of indicators is that it could be used at district level to broadly generalize the nature of the problem, but it could also be used more carefully at village level to identify local problem areas.

We can give a concrete demonstration of how this process works by looking at one of the SADCC states in detail, using a combination of the methods outlined.

MALAWI: A COUNTRY EXAMPLE
(see map on p. 103)

We begin by analysing the country according to the population density and woody-biomass availability. It is then important to evaluate trends of change in the worst problem areas of high population density and low woody biomass. Obtaining a dynamic appreciation of the situation is vital. Next we examine the effects of the production systems on woody biomass and the relationship of fuelwood to the total resource base of the areas identified.

Low population density and extensive woody biomass

In Malawi, these areas are found in the far north of the country on the Tanzanian border; the hilly (escarpment) region along the shores of Lake Malawi in the north and north-central parts of the country; the mountainous Kirk Range on the Malawi–Mozambique (Tete Province) border between Ntcheu and Mwanza; and also in other, smaller, scattered pockets of land.

In such areas there is generally a low level of development compared to other parts of Malawi. Commercially orientated agriculture is restricted to maize and there is a lack of infrastructural development and service provision. Consequently, commercial penetration is limited and, because of the mountainous terrain, land clearance is also limited. In such areas the low demand is readily met by local, indigenous woodland and the lack of capital available restricts switching to commercial fuels. Such areas represent the best scenario in Malawi, and are under no immediate threat in terms of fuelwood security.

Low population density and low-level woody biomass

Few areas in Malawi fall into this category. The most extensive location is the area to the north of Kasugu and to the west of the Viphya Mountains and the Buranje Valley.

If it were not for the low population density, and therefore low fuelwood demand, these would .be significant areas of critical shortage. As it is, the main difference between these and the better-wooded areas with similar population densities is the increased time allocated to fuelwood collection. Rapid population increase could make these into areas of resource scarcity within a few years, as has happened in other areas of Malawi. Development plans for such areas need to recognize their limited resource potential and any interventions will need to take care to maintain current levels of woody-biomass production.

Difficulties could rapidly arise in the Buranje valley when, as a response to the breakdown of the agricultural production system in the mountains, Mozambican refugees cross the border from Tete into the area between Dedza and Ntcheu and then filter eastward down into the valley.

High population density and extensive woody biomass

Two small but significant pockets of high population density exist in the extensive indigenous woodlands of northern Malawi around Karonga and the Mzuzu-Nkata Bay axis (the latter being an area of moderate recent population increase). In the Karonga area, there is a moderately high rural population density, particulaly around the shores of the lake. Cultivation density is high, with extensive cotton and rice production. Along the Mzuzu-Nkata Bay axis there is a dual problem: a higher than average rural population and cultivation density (crops include rice, cotton and rubber) and the urbanization of Mzuzu.

There is no immediate shortage of fuelwood in this area. However, medium- and long-term projections for these two areas, especially the Mzuzu-Nkata Bay axis, reveal a declining fuelwood resource as land is cleared for cash crops and urban and rural centre demand grows, strongly influenced by service and infrastructural provision. This growth in demand will lead to more extensive exploitation of the indigenous woodland along new roads. The Mzuzu-Karonga--Tanzania road link will be especially important as increased reliance is placed on Dar-es-Salaam as an entrepôt for Malawi.

High population density and low-level woody biomass

These areas represent the worst scenario in terms of fuelwood supply. In Malawi, seven such locations can be identified:

• Dedza-Ntcheu Highland Region
• Lilongwe Plain
• Phalombe Plains, north of Mount Mulanje
• Shire Valley
• southernmost part of the shore of Lake Malawi and Lake Malombe
• area of tea plantations centred on Thyolo
• Likoma and Chizumulu Islands.

These areas display a number of common characteristics and the variations in these can be used to divide them into groups. The three main variables are population, agricultural system and fuelwood supply. We will examine each of these in turn.

(a) The first variable is population. In one category, population density is high but perhaps only due to a rapid and recent increase in rural areas (for example, the Lilongwe Plain, the middle Shire Valley and the Dedza-Ntcheu region).

Less rapidly increasing populations are found in the Phalombe Plains, the northern and parts of the central Shire valley, much of the Lake Malawi and Lake Malombe shoreline, and the tea plantations around Thyolo.

Decreasing rural populations are found in the southern Shire valley, and the Likoma and Chizumulu Islands.

(b) Commercial agriculture has a strong influence on fuelwood collection which varies according to different levels of capital flow.

Estates alienate the rural population from land and biomass resources both on and around them. This is an important factor in the tobacco-growing regions of the Lilongwe Plain, middle Shire Valley and Phalombe Plains as well as in the Thyolo–Mulanje area where tea, coffee and tobacco are major crops.

Small-scale vegetable cash-crop production for roadside selling is well-developed in the Dedza-Ntcheu Highland region. Fertile valley land is given over to vegetable production and labour-time is taken up in this selling. The capital returns from this *should* offset the lost fuelwood-collection time (as both tasks are being done by women) by enabling them to engage in intra-rural fuelwood purchasing *if* the system is sustainable on a long-term basis. If not, it indicates that the system is breaking down.

The shores and the islands of Lake Malawi do not have any commercial agriculture. However, it can be argued that these areas may be equivalent to the Dedza–Ntcheu Highlands in that roadside and commercial fish-sales are economically important.

(c) The relationship of fuelwood to the total resource base of the regions is an important holistic concept which has implications for both domestic labour and capital-budgeting, and the long-term sustainability of the systems. A rating of the seven regions can be suggested (Figure 6.1) if the following resources are also considered: water availability; soil fertility (including soil erosion); land fragmentation; and energy resources.

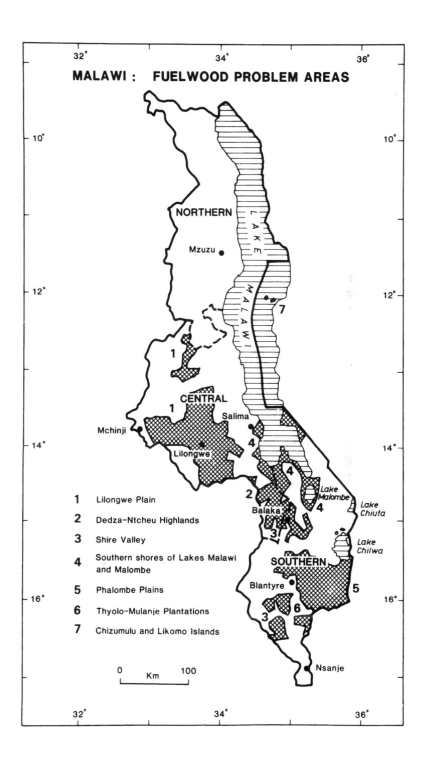

MALAWI :   FUELWOOD PROBLEM AREAS

32°   34°   36°

10°   10°

NORTHERN

Mzuzu

12°   12°

7

1

CENTRAL

1

Salima

Mchinji

14°   Lilongwe   14°

4

2   Lake
Malombe

4

1 Lilongwe Plain   Balaka   Lake
Chiuta

2 Dedza-Ntcheu Highlands   4

3 Shire Valley   3   Lake
Chilwa

4 Southern shores of Lakes Malawi   SOUTHERN
and Malombe

5 Phalombe Plains   Blantyre   5

16°   16°

6 Thyolo-Mulanje Plantations   6

3

7 Chizumulu and Likomo Islands

0   100
Km

Nsanje

32°   34°   36°

Table 6.1 Malawi's Fuelwood Problem Areas

Region	Land Fragmentation	Soil Fertility	Water Availability	Fuelwood Supply	Current Situation and Long-term Prognosis
Lilongwe Plain	** (incl. alienation)	* (in parts **)	**	**	Very restricted resource base, will become more serious.
Dedza–Ntcheu Region	***	***	*	*** (decreasing)	Extremely restricted resource base, problems will increase.
Phalombe Plains	**	*	**/***	**	Very restricted resource base, will show slight increase in seriousness.
Shire Valley	**	*	**/***	**	Very restricted resource base, will show slight to moderate increase in seriousness.
S. Lake Malawi and Lake Malombe	**	*	–/*	**	Restricted resource base, will increase to very restricted in some areas.
Plantation area Thyolo	*** mainly land alienation	**	–	**	Restricted to very restricted resource base, will increase rapidly.
Likomo and Chizumulu	***	***	***	***	Extremely restricted resource base.

Degree of Problem: extreme – ***, very high – **, high – *, lower absent –

If we consider all three variables together, the problem areas can be divided into the following four categories:

1. Those with an extremely restricted resource base, a rapidly increasing population and non-estate commercial agriculture: the Dedza–Ntcheu Highlands.

2. Those with a restricted or severely restricted resource base, an increasing population (sometimes quite rapidly) and estate or non-estate commercial agriculture: Lilongwe Plain, Phalombo Plain, middle and northern Shire Valley, and the Thyolo region.

3. Those with an increasingly restricted resource base, a growing population and little or no commercial agriculture (but fishing as a substitute): the shores of Lake Malawi and Lake Malombe.

4. Those with a severely restricted or even an extremely restricted resource base, little or no commercial agriculture and a declining population: the Lower Shire Valley, and the Likoma and Chimuzulu Islands.

In arriving at a crude identification of problem areas, the first two impact parameters of the fuelwood situation to consider, are the extent of the woody-biomass resource and demand according to population density. We began the case study by isolating four broad categories according to these impact parameters (see Table 6.2) and then provided a more detailed analysis which brought in other major impact parameters, focusing on a single SADCC country, Malawi.

Table 6.2 Population Density and the Woody-Biomass Resource Base

	Extensive Woody Biomass	*Low Woody Biomass*
High Population	High population and extensive woody biomass	High population and low-level woody biomass
Low Population	Low population and extensive woody biomass	Low population and low-level woody biomass

The processes outlined can stimulate innovation, particularly in areas of high population and intensive production. Subsequent interventions can usefully build upon existing developments in improved woody-biomass management. The important lesson to be learned is that the wider processes which have an impact upon the availability of woody biomass must be understood if effective intervention is to occur.

REFERENCES

1. M.R. de Montalembert and J. Clement, *Fuelwood Supplies in Developing Countries*, FAO Forestry Paper No. 42 (Rome: FAO, 1983).
2. G. Foley, *Discussion Paper on Demand Management*, World Bank Eastern and Southern African Regional seminar on household energy planning (Harare: February 1988), p. 6.
3. See A. Millington *et. al.*, *Biomass Assessment* (London: Earthscan Publications, 1988).
4. G. Leach and M. Gowen, *Household Energy Handbook*, World Bank Technical Paper No. 67 (Washington DC: World Bank, 1987), p. 128.
5. For the detailed picture see A. Millington *et al.*, op. cit.

PART THREE

Urban Areas

7. Urban Energy Use

The fuelwood problem in the cities and towns of the SADCC region is qualitatively different from that experienced in the rural areas. The fuelwood trap is still there, waiting to catch the unwary and it takes a variety of forms. First there is the temptation to assume that urbanization will automatically produce new patterns of energy consumption with modern fuels replacing traditional ones, hence no need for intervention. Higher cash earnings and urban lifestyles will lead consumers to purchase cleaner, more efficient, technologically superior and convenient sources of energy such as electricity, gas, coal and kerosene. Urbanization is the cutting edge of economic development and evidence to date indicates that it is associated worldwide with the transition to modern energy sources. Leach and Gowen point out that biofuels account for 60–95 per cent of total energy use in the poorest countries, 25–60 per cent in the middle-income countries, and less than 5 per cent in the high-income countries.[1]

If fuelwood continues to remain in demand during the transition to modernity and there is a shortage, then peri-urban fuelwood plantations are frequently proposed as a solution to meet the shortfall. Some argue that the fuelwood problem experienced particularly in the urban areas, could be resolved fairly easily by promoting a switch to kerosene, a move one rung up the energy ladder.[2] As kerosene accounts for only a small percentage of total oil imports, according to this argument there would not be a significant cost-effect on the balance of payments.

Whilst not dismissing the validity of the argument linking economic development with an energy transition away from fuelwood, nor the need in certain circumstances to encourage peri-urban plantations or a massive fuelswitch to kerosene, each carries within it the seeds of failure when unthinkingly applied as policy.

The problem with an approach which believes that modernization is the solution, is the basic assumption of continuing economic growth which exceeds not only the rise in population but also the much higher rate of urban growth. In other words, such an approach ignores the current reality in the majority of SADCC member states of declining per capita income, negative rates of growth, rising debt and an extreme scarcity of foreign exchange. According to the World Bank, per capita income declined 12 per cent over the first half of the decade in the low-income African countries.[3] It also declined for both the middle-income oil-importing and oil-exporting countries.[4] There is every indication that this downward trend may continue into the next decade and possibly beyond for some, though not all, of the SADCC member states.

The difficulty with assuming that peri-urban plantations can meet the fuelwood needs of urban centres is that the resulting cost of production is generally far higher than consumers can afford or are willing to pay, necessitating a heavy state subsidy. As urban centres grow, the opportunity cost of the land increases and generally the rate of return on food production or building land will vastly exceed that of fuelwood production.

Kerosene may appear to be a natural substitute for fuelwood. But even if the increase in the total foreign exchange bill might not be so great, in a period of foreign exchange shortages governments are unlikely to take this option as it will mean subsidizing consumption rather than production.

These then are just a few of the traps which await those trying to wish away or "solve" the urban fuelwood problem. As with the rural fuelwood problem, the first task in the urban areas is to conceptualize the problem properly, understanding both its complexity and specificity from one location to the next. This is difficult because, strange as it might seem, we know less about the urban fuelwood problem and how to tackle it than we do about the rural problem. There is a real and pressing need to acquire a better and more comprehensive data-base upon which to construct an analysis. Without it, conclusions at this point in time must remain tentative.

We will begin by examining the urban context of energy use and then go on to consider the various factors that shape energy consumption. To get to the root of the problem it is essential to ask the right questions. Specifically, why are certain fuel choices made? Is it simply a question of consumers comparing relative prices for units of energy delivered? Is it basically income levels which determine whether cash is available for the initial high-capital investment required, say, for a household to "go electric"? Or are other factors such as security of supply the dominant concern, given the prevailing conditions in many countries of the region?

URBAN ENERGY CONTEXT

The countries of southern Africa are all experiencing a period of rapid urban growth (see Table 7.1) with a parallel rise in urban energy consumption. In the likely event that present trends continue, urban energy-consumption may match or even exceed rural consumption within 20 years in most SADCC states. Fuelwood is an important component of urban energy-consumption, particularly in the domestic sphere. It is the poor in urban societies who mainly rely on firewood or charcoal and meeting their energy needs should be a concern of both governments and aid agencies.

The process of urbanization reflects the ever greater integration of developing countries into the global economy. Cities are the channel through which this process occurs. It is generally accepted that there is an urban bias in the development process which creates widening gaps in income, access to services and opportunities for advancement, between the rural and urban areas.[5] Economic and social power is concentrated in the cities which develop whilst rural areas stagnate. The widening rural–urban gap encourages further migration from the countryside and the towns grow even faster. Demographic trends and historical evidence from other locations lend support to the prediction that the SADCC region will become substantially urbanized during the next century.

Urbanization is an uneven process. It is the large cities and the capitals in particular, which grow the fastest; a feature known as urban "primacy". This leads to an irregular spread of ubanization, potentially creating acute pressures on the woody-biomass resource base.

Whilst the fuelwood problem is different in the rural and urban areas, the interconnections between the two should never be forgotten. As urban centres mushroom, the search for fuelwood and charcoal to feed the rising demand begins to denude the surrounding countryside of trees. The cost of producing charcoal, which is easier to transport for urban needs, is a far higher rate of tree loss. As urban centres grow, the zone of exploitation increases. The supply zone around the towns and cities grows and takes on an amoeba shape as the highways feeding the demand become conveyor-belts for moving fuelwood to the towns.[6] Urban demand for fuel is a major cause of environmental degradation in the countryside.

Table 7.1 Urbanization in SADCC Countries

	Urban pop. as % of total pop.		*Average annual % pop. change* 1975–80		*1985–90*	
	1960	1985	urban	rural	urban	rural
Angola	10·4	24·5	6·7	2·6	5·5	1·6
Botswana	1·8	19·2	8·5	3·0	7·7	2·5
Lesotho	1·5	5·8	7·6	1·9	7·1	2·3
Malawi	4·4	12·0	7·6	2·4	7·4	2·7
Mozambique	3·7	19·4	12·8	3·4	9·2	0·9
Swaziland	3·9	26·3	9·8	1·5	7·7	1·2
Tanzania	4·8	14·8	8·5	2·8	7·7	2·9
Zambia	17·2	49·5	6·3	1·0	5·8	0·8
Zimbabwe	12·6	24·6	5·7	2·8	5·9	2·8

NOTE: Country comparisons are imprecise owing to different definitions for the size of urban versus rural settlements.

SOURCE: World Resources Institute and IIED, 1987. *World Resources 1987*, based on data from UN Population Division. Adapted from Leach and Mearns (1988).

WHAT SHAPES ENERGY CONSUMPTION?

The extent of fuelwood use differs from one urban location to the next and no accurate and comprehensive data exists on urban energy-consumption for any SADCC country, let alone a comparative set of statistics for them all. Therefore, the task of understanding the problem is especially difficult. We have to tease out the evidence from the limited data available and are necessarily obliged to draw our conclusions with caution. Although the pattern of urban energy-consumption will differ from place to place depending upon the particular combination of environmental and socio-economic factors involved, there are certain general observations that can be made. However, each of these generalizations may be open to contradiction for any specific case.

We will now try to indicate some of the more important factors that shape the patterns of urban domestic energy-consumption. The first of these is the size of the town, which will influence the types of energy consumed. Broadly speaking, the smaller towns will reflect more traditional, rural patterns of energy consumption, relying heavily upon firewood and charcoal (where the latter is used) for domestic energy purposes. As the urban centres grow, the availability of commercial energy sources increases and firewood is more likely to be replaced by other fuels. Not only will the percentage of households using other fuels increase but the quantity of fuel consumed per household is likely to be greater, at least in the early transition stage. Table 7.2 provides some supporting evidence in the case of Zambia.

Table 7.2 Variations in Urban Fuel Use in Zambia

Zambia (per household consumption 1980)

	Large Urban Centres		Small Urban Centres	
	% using	consumption (GJ*)	% using	consumption (GJ)
Fuelwood	26	42	65	75
Charcoal	87	29	64	16·5
Kerosene	76	2·1	77	1·7

SOURCE: ETC, *SADCC Fuelwood Study*, Leusden, 1987.
* GJ = Giga-joules

A second generalization is that a distinction exists between the energy mix used in the wealthier, low population density suburbs of a city or town compared with the higher population density low-income areas. These poorer areas are known variously as locations, shantytowns, or "musseques" and "canico" suburbs in the case of the ex-Portuguese colonies of Angola and Mozambique. Table 7.3 provides the available evidence to support this in the case of Zimbabwe. Consumers in the low population density suburbs account for 70 per cent of all the electricity and liquefied petroleum gas (LPG) consumed. Those living in high population density suburbs account for the bulk of fuelwood and paraffin consumption. It is interesting to note from the table that households in the areas of low population density consume almost twice the energy of the high population density suburbs. So, a crucial distinguishing feature of each location is not simply the mix of energy used but also the amount of energy consumed. In particular, those households with electricity in the lower population density suburbs have a greater variety of uses for energy in the form of consumer goods to enhance and modernize their quality of life. These take the form of radios, record-players, televisions, videos, air conditioning, food mixers, personal computers and the like. Quite simply, as the quality of modern urban life improves domestic energy-consumption increases.

If we take the issue of urban social differentiation further then

Table 7.3 Patterns of Domestic Energy Consumption in Urban Areas

ZIMBABWE   (Total urban energy-consumption in PJ)

	High-Density Areas (67% urban pop.)		Low-Density Areas (33% urban pop.)	
	PJ	*% Total*	*PJ*	*% Total*
Kerosene	0·15	4	0·1	3
LPG	0·01	1	0·01	1
Electricity	0·88	22	2·57	73
Coal	0·07	17	–	–
Fuelwood	2·35	57	0·84	24
TOTAL	4·09		3·52	

MALAWI   (Main fuels used for cooking by income group, % of total)

	High Income	Middle Income	Low Income	TOTAL
Wood	8	25	75	68
Charcoal	11	47	21	22
Electricity	80	26	2	7
Other	2	2	2	2
Total urban pop.	12	24	64	100
Household Fuel Expenditure (Kwacha/month)	53·2	26·4	9·6	12·2

SOURCE: ETC, *SADCC Fuelwood Study*, Leusden, 1987.

again a distinctive pattern of consumption emerges. This is demonstrated in the case of Malawi in Table 7.3. Low-income households rely upon wood for cooking, while high-income households use electricity. Evidence from the SADCC region, surrounding countries and from further afield suggests that charcoal, kerosene and LPG are the fuels most commonly found in the transition between wood and electricity. The extent of energy consumption for non-cooking purposes also increases with income, notably the provision of artificial light and audio-visual entertainment. While income and location may act to the same effect, this is not always the case. For example, the figures for household energy consumption in low population density housing in Zimbabwe in Table 7.3 include both low-paid servants and their high-income employers. Their location may give the servants access to electricity, yet their earnings may be less than those of a sizeable number of inhabitants of the townships.

Urban energy-consumption will naturally vary according to the size of the household. Many single, male migrant workers in the cities may not do a lot of cooking in their homes, relying instead on cooked food purchased in the market or on the roadside at convenient locations. As one recent study has observed, "Household size often has as great or greater effect on energy consumption as other major variables such as income."[7]

The number of influences determining the pattern of energy consumption in a given urban area vary and the impact of the different forces will inevitably produce a situation unique to each location. This pattern can also change quite rapidly in any one place if there are changes in fuel availability, fuel prices or income. The dynamic of the energy transition leads generally from biomass fuels towards industrially produced energy sources. Yet this transition can also be reversed, with backward fuelswitching. As we shall see, a variety of strategies are being employed to alleviate the difficult circumstances facing many people in the region.

WHY ARE FUEL CHOICES MADE?

In urban centres, fuelwood is no longer a freely available good. It is one more energy commodity to be sold on the market and it competes with a whole array of other fuels. This might lead one to the inevitable conclusion that price is the only relevant factor in determining choice. This, however, is not the case. Indeed, it is one more trap to snare the unwary. As we will demonstrate, while price is very important and is a complex issue, risk-minimization and the ensured stability of supply are the critical concerns for most urban households. This means that the availability and reliability of supply are of central importance, given the prevailing conditions within Southern Africa. Naturally, cultural preferences will also play their part. Whilst little work has been carried out upon these preferences, this does not diminish their importance.

The various fuels available have different properties, which may make them more or less attractive to the individual consumer. Electricity is clean, efficient, instantly available and can simmer or boil food at the turn of a dial. It also involves no time spent on transporting supplies from the market and is permanently on tap inside the home. Bottled gas has many of the same properties as electricity for purposes of cooking but does not have its versatility in providing lighting, air conditioning, heating and entertainment (via radio, video, hi-fi and televison). Gas also involves collecting heavy bottles, which requires some form of transport.

Kerosene is a good fuel for quick water-boiling but the cooking of main meals can take too long and the smallness of the stove may make it unstable given the size and shape of traditional cooking utensils. This may create a domestic hazard with a rapidly rising population resulting in an abundance of small children around in the kitchen. The smell of kerosene can also deter people.

Charcoal is easier to transport than fuelwood, both are smokey but they also give a certain taste to food which many people like, or at least have become accustomed to, and is now a cultural preference. Both wood and charcoal provide more heat than kerosene. These variable properties mean that not all fuels are direct substitutes for each other.

Why do people make the fuel choices that they do? We have to confess that the hard evidence upon which to base a meaningful assessment is just not available. All that we can hope to do at this stage is explore the issue as thoroughly as possible. There is an urgent need for more action-orientated research. The obvious answer is to apply basic economic calculations to the problem. Yet, as we shall argue, this may not explain the existing situation faced by many people within the region which may determine a different causal logic.

Let us begin with the orthodox analysis of comparative costs. The calculation of these costs is not an easy matter. As we will soon

observe, financial comparisons alone, are insufficient, as consideration of the *relative efficiency* of any combination of fuel devices is vital
and these can be difficult to assess.

The great advantage of fuelwood, charcoal and kerosene to the
urban poor is that they can be purchased in small quantities on a
daily basis. Scarce cash resources do not permit the luxury of large,
one-off financial outlays, even with the eventual promise of cheaper
power.

Cost is not just the price of fuel but also the cost of the necessary
appliances, which may be substantial and which must generally be
paid in a lump sum – a "capital" cost. Access to sufficient cash to pay
for expensive appliances is a real problem for many of the urban
poor. The manner in which fuels are paid for is also important.
Electricity necessitates the ability to save for monthly or quarterly
bills, LPG to pay for a bottle of gas all at once, and so on. Many of
the urban poor do not have the disposable income to accumulate
these lump sums, even if their fuel costs would be cheaper in the long
run. There is also the cost of maintaining the supply system.
Furthermore, in many places electricity for domestic use is dependent upon certain standards of housing or a particular location in the
city. These and other factors mean that the *cost* of different fuels
consist of more than just the *price* of that fuel. The key issue is the
way in which these costs relate to the pattern of life for the urban
poor.

Fuelwood in urban areas differs from other fuels in its nature as a
commodity. In particular, it is produced in the petty-commodity
sector by numerous producers from a range of different sources of
supply. In contrast, commercial fuels are frequently imported, and
are produced and distributed by the modern market sector or the
state. This means that the cost and scale of production of fuelwood
is far more flexible than those of commercial fuels, and that many
petty producers will continue to extract, transport and sell fuelwood,
whatever the price. They frequently have little choice in the form of
available alternative employment.

The evidence available from within the SADCC region, patchy as
it may be, indicates that the availability and security of the supply of
energy is frequently a dominant consideration in consumer choice.
There are many and varied circumstances contributing to the
insecurity and lack of availability of sources of energy supply.

Angola and Mozambique bear the brunt of a bloody war of
destabilization engineered from Pretoria. The provision of energy

supplies is highly precarious with pylons, power stations and oil storage depots being targets of attack. Many more countries have had their energy supplies interfered with either because they are imported directly from South Africa or because they rely on South African transport routes for imports to reach them in the interior of the sub-continent. The South African government frequently uses this dependency to apply pressure, by interrupting sources of supply. In addition to destabilization, poor road and rail routes add to the problem of reliability. Meals still need to be cooked when domestic energy supplies are interrupted.

World recession, declining terms of trade and rising debt are severely limiting foreign exchange for the countries of the SADCC region. Increasing the provision of industrially produced fuels usually requires a heavy foreign-exchange expenditure and this is frequently not available. Using scarce foreign exchange for domestic-energy provision has not normally been high on a government's list of priorities. In the case of electricity, everything from the materials for building power plants down to domestic electrical appliances is generally imported and spare parts prove extremely difficult to acquire. As more and more countries in sub-Saharan Africa are obliged to adopt structural adjustment programmes with the accompanying devaluation of their exchange rates, the cost of such imports increases considerably.

Like the rural subsistence farmers, urban energy consumers adopt a strategy of *risk minimization*. This takes various forms. The first concerns those options open to the poorer sectors of the urban community. They tend to stick to fuelwood or charcoal for major domestic energy needs, notably the cooking of main meals and heating in winter.

The open fire is very versatile and provides a family focus. It is a psychological link with life back in the rural areas when news was exchanged, the lineage history transmitted, the religious cosmology explained and the wisdom within the farming community shared, over the flames and embers of the open fire. When the daily work was done, signalled by the rapid disappearance of the tropical sun between five and six in the evening, the log fire became the focus for family and community alike. Dreams were spun while looking at the dancing flames and the art of conversation flourished. The electrical imagery and sound of radios, record-players, videos and televisions have irrevocably destroyed that other era.

REFERENCES

1. G. Leach and M. Gowen, *Household Energy Handbook*, World Bank Technical Paper No. 67 (Washington DC: World Bank, 1987).
2. G. Foley, "Woodfuel and conventional fuel demands in the developing world", *Ambio*, Vol. XIV, Nos. 4–5, 1985.
3. World Bank, *Financing Adjustment with Growth in Sub-Saharan Africa, 1986–90* (Washington DC: World Bank, 1986), p. 9.
4. World Bank, *Towards Sustainable Development in Sub-Saharan Africa* (Washington DC: World Bank, 1984), p. 10.
5. See, for example, M. Lipton, "Why poor people stay poor", in J. Harriss (ed), *Rural Development* (London: Hutchinson University Library for Africa, 1982).
6. J. McClintock, *Fuelwood Scarcity in Rural Africa: Possible Remedies*, Vol. 1 (Centre for Agricultural Strategy, University of Reading, 1987), pp. 19–20.
7. G. Leach and M. Gowen, op. cit., p. 47.

8. *Urban Energy Options*

Given rapidly rising rates of urbanization, deforestation and foreign trade imbalances, it is vital to begin facing the problems of how to ensure urban energy supply. As we have seen, little reliable information exists and there is an urgent need for a more detailed investigation of urban energy demand and supply, patterns of use, and price movements over time.

While reviewing the urban energy situation, one striking aspect that emerged was the way in which supply reliability determines the pattern of domestic energy use. This is a feature of all income groups, whatever the energy mix used in their homes. In many places there is as likely to be a shortage of gas bottles, as an interruption in electricity supply. The rich, of course, have the luxury of being able to afford to keep a wide array of options open, they simply double-up on energy appliances and supplies.

Urban energy-consumption patterns have a tremendous impact upon the rural areas, both directly and indirectly. As urban populations increase, more land has to be cleared to feed the growing band of urban non-subsistence producers. The demand for firewood and charcoal inevitably increases. In essence, the site of fuelwood consumption grows at the expense of areas of fuelwood production. This has a negative effect upon the environment as deforestation accelerates in the growing urban hinterland, particularly along the major arterial routes. The richer families switch away from bio-energy sources (moving up the energy ladder to kerosene, gas and electricity), and this may ease the pressure upon the rural environment. However, this almost inevitably leads to a greater demand on scarce foreign exchange resources and can thereby contribute to the wider bias towards urban development to the cost of the rural areas.

Increasing energy use is necessary for economic growth and urban development. Yet, given the circumstances in southern Africa, it is important to try to maintain the security of energy supply which implies guaranteeing a measure of choice. Disruptions in domestic and commercial energy supplies can lead to a fall in production: the health and vitality of workers may be affected by the former and production runs interrupted in the case of the latter. Given the pressures of international competition, it is also imperative to find the cheapest sources of supply. High energy costs make the problems of obtaining improved industrial efficiency that much greater, and for the poor it is an additional heavy burden.

What possibilities exist for tackling the urban fuelwood problem? Essentially, there are four viable options: fuelswitching, supply enhancement, and (of lesser importance) conservation and improvements in transport and marketing. We will examine each in turn.

FUELSWITCHING

Urbanization creates the context for a move away from fuelwood as a source of domestic energy. All the available evidence suggests that cities are, in the words of Leach and Mearns, "the demand-driven growth centres for modern fuel consumption. Combined with their political and economic dominance and favoured position over infrastructure investments, they are also best served by modern fuel supply and distribution systems."[1]

The larger urban areas appear to possess two of the most important forces propelling the domestic energy transition away from biofuels: the access to alternative energy supplies and higher incomes which provide the initial capital outlay to purchase the more expensive devices associated with moving up the energy ladder.

Fuel substitution is one important way of mitigating the growth of fuelwood demand in the urban centres. To be effective, however, it needs to follow on-going trends rather than fly in the face of them. It is always easier to facilitate fuelswitching in the direction to which consumers' preferences are already moving rather than attempt to reverse market trends.

One important criterion in determining the prospects for fuelswitching is the comparative costs. In Table 8.1, we have compared the relative running costs for various fuel/device combinations in the major SADCC cities. This is based upon taking the local retail price of firewood and the use of an open or three-stone fire, as the numeraire. Great care has to be taken in interpreting these figures, which are only indicative, as certain assumptions are made concern-

Table 8.1 Relative Running Costs for Various Fuel/Device Combinations in Major SADCC Cities

	Gaberone	Harare	Lusaka	Mbabane	Maseru	Maputo	Luanda	Lilongwe	Dar es Salaam
3-Stone Fire/Firewood	1·00	1·00	1·00	1·00	1·00	1·00	NA	1·00	1·00
Improved Fire/Firewood	0·53	0·50	–	–	0·56	–	NA	0·50	–
Traditional Stove/Charcoal									
Retail	–	–	1·17	–	–	6·30	NA	1·61	1·49
Wholesale	–	–	–	–	–	1·25	NA	–	–
Improved Stove/Charcoal									
Retail	–	–	0·80	–	–	–	NA	–	0·83
Wholesale	–	–	–	–	–	–	NA	–	–
Coal Stove	0·06	0·16	–	0·33	0·47	–	NA	–	–
Wick Stove/Kerosene	0·54	0·60	0·69	1·34	0·61	?	NA	1·33	1.64
Gas Stove/LPG	0·71	–	–	–	–	–	NA	–	–
Hot Plate/Electricity	0·94	0·28	0·21	0·68	?	?	NA	1·26	–
Cooker/Electricity	–	0·43	0·17	–	?	–	NA	–	0·9-10

Ratios taking the local retail price of firewood and the use of an open fireplace or 3-stone fire as the numeraire

*1 At unofficial price of 1·16.
*2 These ratios are based on 1983 prices.
*3 With bucket stove.

ing relative efficiencies in the use of the fuels. Also, the data source for the region is weak.

The results make interesting, if potentially deceptive, reading. Coal stoves appear by far the cheapest option for those countries possessing coal resources or having easy access to them. Yet they remain relatively uncommon. Electricity comes a close second, at least in Harare and Lusaka which reap the advantages of cheap hydro-electric power from Kariba and, for Zimbabwe, the abundant coal reserves at Wankie which are used for thermal production. Yet electricity remains available only to a small minority. Use of kerosene with a wick stove is more common and this appears to be a cheaper option to firewood at least in Gaberone, Harare, Lusaka and Maseru. Elsewhere it compares unfavourably with the traditional open fire. Both the improved firewood and improved charcoal stoves appear to produce considerable cost savings.

In Table 8.2, we try to refine the analysis further. This is a more accurate comparison of costs because it takes into account the discounted cost of the device employed. It also shows the complete breakdown of the way in which such figures are produced for one SADCC city, Lusaka.

Of course, much of what we imply from this table depends upon the accuracy of estimates of energy efficiency and these are only assumptions, no more than this. Whilst there are major cost savings with electricity, the high initial capital expenditure restricts this option to a small number of high-income earners. Looking carefully at the figures, one sees that major savings are experienced with improved firewood, charcoal and kerosene stoves. Hence, there appears to be a strong link between conservation and fuelswitching in that conservation can dramatically alter cost calculations. The retail cost of firewood, charcoal and kerosene for unimproved devices does not differ significantly.

Taken together, and ever mindful of reservations about the quality of the data, the two tables show apparently surprising results, notably that fuelwood is relatively expensive in many SADCC cities. It is particularly expensive when compared with coal or hydro-electric power produced from local rather than imported sources. (However, it should be noted that it appears cheaper than charcoal or kerosene in Lusaka when comparing fuel costs involving improved stoves.) On the positive side, this suggests that the possibilities for fuelswitching looking promising. This may make it possible to ease the pressure upon the wood resource

Table 8.2 Equivalent Costs of Different Device/Fuel Combinations in Lusaka
(using cost and price data from mid 1986)

Device and Fuel Used	Efficiency of Device %	Combustion Heat Reqs mJ	Heat Value of Fuel mJ/Kg	Weight of Fuel needed per month Kg		Cost of Fuel Zk/Kg	Running cost per Month Zk	Life of Device Months	Purchase cost of Device Zk	Overall (average) cost/mth Zk
3-Stone Fire/ Firewood	10	4750	18	264	Wholesale Retail	0·02 0·27	5·28 71·28	–	0·00	5·28 71·28
Improved Stove/ Firewood	20	2375	18	132	Retail	0·27	35·64	12?	?	?
Mbaula/Charcoal	15	3167	30	106	Wholesale Retail	0·32 0·79	33·92 83·74	15	5·00	34·25 84·07
Improved Unza Stove/Charcoal	22	2159	30	72	Wholesale Retail	0·32 0·79	23·04 56·88	36	25·00	23·73 57·57
Wick Burner/ Kerosene	37	1284	42	31	Official Retail	1·59 2·67	49·29 82·77	36	36·00	50·29 83·77
Pressure Stove/ Kerosene	50	950	42	23	Official Retail	1·59 2·67	36·57 61·41	60	170·00	39·40 64·24
	%	kWh		kWh		Zk/kWh	Zk	Months	Zk	Zk
Single Hot Plate/ Electricity	62	210	–	210		0·07	14·70	72	285·00	21·16
Cooker/Oven/ Electricity	75	176	–	176		0·07	12·32	96	1990·00	35·55

without placing an undue burden upon tight government budgets. A major problem is that we know too little about price trends in African cities for fuelwood and its substitutes. Leach and Mearns have pulled together such reasonably reliable information as exists, but no clear pattern emerges.[2] They look to the model produced by Douglas Barnes at the World Bank to try to explain the relationship between declining fuelwood resources and fuelwood prices.[3] According to this model there is a first stage in which fuelwood is readily available at a low price. Low population densities do not place a heavy demand upon the resource and fuelwood becomes available from land clearance. In a second stage, growing population pressures make their impact felt. Wood comes from land clearance along the main arterial routes over ever greater distances and transport costs rise. In some instances there is a switch to charcoal which may reduce transport costs but carries a heavy burden in terms of the increased fuelwood consumption necessary for its production. It is now that prices might rise, sometimes in spectacular fashion, but the volatility of this makes it risky to extrapolate price trends.

In stage three, the pressure on forest resources is severe and fuelwood prices are high. However, other energy sources are available and these place a ceiling on fuelwood prices with the fuelwood trade having to adjust its costs accordingly. Prices then tend to fluctuate according to the relative costs of alternative fuels. Leach and Mearns conclude that: "Rising income seems to be the main driving force for switching from fuelwoods into other fuels, and not high fuelwood prices compared to competing fuels."[4]

If it were merely a matter of cost factors determining consumer preference then why do people continue to rely on fuelwood or charcoal? The answer would appear to lie in other considerations. Poorer people may not have either the capital available or a house in a condition or location good enough for the transition to more modern forms of energy. The evidence appears to support our central argument that it is the availability of fuels and security of access, rather than comparative costs, which may impair a more rapid fuelswitching transition.

In SADCC countries experiencing the effects of war and economic recession, fuelswitching back to fuelwood is common, as the reliability of supply is generally greater than that of alternatives. A survey in 1987, in Dar es Salaam, Tanzania, discovered that of those households switching fuels, a majority had moved back down the energy ladder to charcoal and fuelwood.[5] This reinforces the point that ensuring supplies of fuelwood is likely to remain important into the foreseeable future, given economic and political uncertainties.

In order to encourage fuelswitching, it is important to guarantee the availability of secure supplies of modern fuels, to ensure that incomes are sufficiently high to facilitate the purchase of more expensive appliances, and to develop appliance supplies at affordable prices. The damaging effects of the world recession, rising debts and the destruction being wrought by South African destabilization in the region, will clearly impose considerable constraints on the achievement of these goals in all locations.

The timeframe is important when considering fuelswitching options. In the short-term, options will be constrained by available supplies, the capacity of the infrastructure, and the existing pattern of fuel consumption. Planning interventions in the longer term is inevitably hazardous. To give one example, the cost of both oil and coal fell dramatically on the world market between 1986 and 1988. Such fluctuations will have a profound impact upon the economic efficiency of any heavy capital investments made in the energy sector.

The fuelswitching policy which appears most attractive is to promote the development of indigenous commercial fuel resources where they are available and go for an urban energy future in which they play a central role. Hence, there appear to be opportunities for coal (for example in Botswana), and hydro-electric power in Zimbabwe and Mozambique. However, a word of caution is needed concerning coal, because a domestic market is only likely to develop following an expansion in the industrial utilization of coal. A fuelswitch approach based on developing indigenous commercial fuel resources, depends upon significantly expanding production and the distribution infrastructure. As such, it is a long-term strategy which will require a careful assessment of the viability of the substantial investments involved. However, perhaps this approach offers the best hope for sustainable future energy supplies where such indigenous resources exist.

Once it is decided which fuels and appliances are to be encouraged, the relevant authorities can take the necessary steps towards ensuring their availability. To be concrete, this will mean providing sufficient kerosene stoves, improved charcoal and fuelwood stoves, LPG cylinders, and cheap electrical appliances. Finding ways of encouraging their production locally is vitally important, with an emphasis on low-cost attractive designs and ease of use. Of further importance is trying to standardize appliances and equipment as this will facilitate the task of improving reliability of supply.

There are a number of practical measures that can be undertaken to encourage the substitution of alternatives to fuelwoods. If the switch is to be to LPG or kerosene, this will imply either changes in the refinery capacity and mix of outputs, or an alteration in import policy. In the case of LPG, a frequent problem is the shortage of cylinders. Efforts can be made to manufacture these locally or within the SADCC region. If it proves possible to produce lower-cost cylinders and appliances, the market could be opened up amongst the lower-income sections of the community. As insecurity of supply is frequently a barrier to increased fuel substitution, greater efforts will be required to improve storage facilities, transportation and the number of retailing outlets.

A good example of the difficulties created by supply insecurity is encountered in the case of LPG distribution in Luanda. The distribution network is focused on the centre of the city, with wholesalers having a number of subsidiaries under contract in the peri-urban areas. The latter give a voucher to the consumer in exchange for empty bottles which entitles the holder to a full bottle when available. Both wholesalers and retailers hold large stocks of bottles, a situation which allows unofficial monopolies and price control. Whilst official prices in 1986 were 180 to 200 kwanzas a bottle, actual prices were as high as 6,000 to 7,000 kwanzas. Households also hoard bottles and many of the bottles are now out of use because they are over 20 years old and worn out. If the government wishes to encourage a greater fuelswitch to LPG, not only will the quantity of gas being produced have to be increased, but so too will the number of bottles. The distribution network will also need to be expanded.

A major advantage of fuelswitching is its flexibility. It can be small- or large-scale and can be phased over time, with different areas of a city included according to fuel availability. Similarly, the particular types of improvements made in different areas can vary to reflect the specific needs and wishes of different target groups. This flexibility is possible because such strategies reflect the ways in which fuels are used in the city. It can also extend to altering the direction of fuelswitches as the relative cost and availability change. For example, an interim measure may be to promote imported kerosene while supplies of indigenous coal, electricity or even wood are developed. When these are available, a policy encouraging a switch to these other fuels can be introduced.

A number of different measures can be employed to facilitate fuelswitching. Improving and extending distribution systems may mean helping to establish government, business or community-run fuel delivery points which could sell a range of fuels and appliances. In certain circumstances, the establishment of a rationing system providing a set amount of fuel for individual households may be required. A campaign promoting the production and distribution of appropriate appliances to target groups can be effective. As far as possible, the strategy should harness the existing marketing systems for fuels. The goal should be to increase secure access for the specified target groups of as wide a range of fuels and appliances as possible.

If such an approach could be successfully established and administered, it has great potential in helping to tackle urban energy problems in the SADCC region as it does not place heavy demands upon financial resources or the planning system. It can be closely targetted and works within the reality of the urban energy systems. Such a strategy would require a very flexible approach and would largely work indirectly. It needs strong community participation as well as the co-operation of the agencies responsible for distributing the various fuels involved. As such, there are administrative hurdles to overcome. If this is achieved, fuelswitching interventions based on improving secure access (rather than subsidizing consumer costs) should be considered as one of the most viable interventions in the cities of many SADCC member states.

Another form of intervention used to encourage fuelswitching is pricing strategies, usually through subsidy and taxation. Pricing strategies have already been widely used within the SADCC region, though not generally in order to encourage fuelswitching. Such interventions can appear superficially attractive given that they require no productive investments and are administratively quite straightforward. For such a policy to encourage fuelswitching without raising prices for the consumer, however, will normally involve subsidizing the fuels targetted for the fuelswitch.

The problems of making such an intervention effective should not be minimized. Basically, there are various *leakage* difficulties. If the price of certain fuels is kept low, they may leak across the border into neighbouring countries where higher prices can be obtained. Kerosene set at too low a price below diesel, will result in the widespread mixing of kerosene and diesel, as happened in Indonesia and elsewhere. Sometimes a different kind of leakage occurs, with consumers switching not away from fuelwood or charcoal but between one or more of the relatively modern fuels.

A somewhat different form of leakage is that of foreign exchange from the government exchequer, which becomes increasingly scarce. For the majority of countries which do not produce the fuel which they wish to encourage the switch to, this immediately requires additional foreign exchange. Kerosene is generally regarded as the bridging fuel for the transition away from dependence upon wood as the source of domestic energy. According to calculations made by Byer, which include three SADCC member states, the increase in total petroleum imports needed in order to replace all urban fuelwood demand with kerosene, would be as low as 11 per cent for Zimbabwe (1980), through to 29 per cent for Malawi (1980) and 74 per cent for Tanzania (1981).[6] Whether these economies could afford this level of subsidy for what is essentially consumption, is another matter. Yet such a move could have profound indirect effects upon the environment and the long-term productivity of the land.

Byer also calculated the effect of the decrease of total demand for fuelwood as a result of such a switch to kerosene. It was limited to a 5 per cent reduction in total demand for Malawi, 18 per cent for Zimbabwe and 29 per cent for Tanzania. To make the necessary calculations about the viability of such a step is extremely complicated. How would employment be affected? Could the economy and debt burden bear the additional cost? Perhaps the most sobering calculation of all is to begin by estimating the realistic rate at which fuel substitution could occur on an annual basis, then simply compare the rate of urban population growth with this figure. In many countries, even a 10 per cent switch to a given fuel would only keep pace with the growth of the urban population. A major fuelswitch initiative may still only tread water, that is maintain the same mix of fuels consumed in urban areas while accommodating ever-growing absolute numbers of people.

The cost of a subsidy-based fuelswitching policy, and particularly the need for foreign exchange, will severely limit the scope for such intervention. Pricing policies are also crucial intervention mechanisms but they tend to lower or raise the prices to all consumers rather than being able to target specific groups. If subsidizing modern fuels is envisaged then it is often more likely to benefit the richer sections of urban society.

What conclusions can be drawn from the knowledge that we possess to date about the options for fuelswitching?

1. The substitution of bioenergy fuels by other energy sources is generally associated with the process of economic and social development.

2. In the specific circumstances in which many people in southern Africa live, guaranteeing the reliability of energy supply and a diversity of available energy options is the key consideration in an environment of uncertain and irregular supply. This is most definitely the case in Angola, Mozambique, Tanzania, Zambia, and elsewhere besides. In terms of the whole SADCC region, these four countries account for the overwhelming majority of the total urban population. Security of supply and availability are the central issues. It is this problem which needs to be tackled first.

South African destabilization has increasingly constrained the arena for development efforts in Mozambique and Angola to the key urban areas. This restriction means that the urban fuelswitching options can become an important developmental opportunity to seize. Yet we must always be mindful of the fact that fuelwood and charcoal will remain of central importance to the poor urban majority into the foreseeable future. For this reason we now turn to the second urban energy option, enhancing fuelwood supply.

SUPPLY ENHANCEMENT

Increasing the supply of fuelwood is an obvious response to the growing demands of the urban centres. It is possible to identify five potential supply sources: peri-urban plantations, natural woodlands, fuelwood from farms, fuelwood as a residue from other production activity, and (in a somewhat different category) agricultural residue used for industrial consumption. Interventions thus far have tended to favour large-scale production on government-run estates. Table 8.3 provides a number of examples from within the region. These peri-urban plantations are generally of exotic species planted in single stands. The peri-urban plantation fits very well within the traditional forestry-department ethos and planning process and thus reflects both the capabilities and the limitations of these. Their appeal is straightforward. The costs and benefits of large-scale plantations are easier to define than many of the indirect approaches put forward in this volume. Their land requirements, personnel needs, capital and income flows and outputs can be costed using traditional forestry-planning skills. Administratively, peri-urban plantations sit comfortably within forestry-department structures and draw upon the store of available skills. They can play a positive role in improving the environment, creating green belts and increasing urban amenities.

Table 8.3 Peri-Urban Plantations: Some Country Examples

MALAWI

1. "Lilongwe City Council Fuelwood Plantations". 3,000 ha. Mainly Gmelina. Begun in 1971 under Capital City Development Authority.
2. "Lilongwe City Fuelwood Plantation Project". 14,000 ha (10,000 ha plantation, 4,000 ha natural woodland). Proposal.
3. "Blantyre City Fuelwood project". 10,000 ha (7,500 mixed eucalyptus, gmelina, pines, melia; 2,500 ha natural woodland). Proposal.
4. 3 "small urban fuelwood and pole plantations". Total 800 ha. Begun in 1980 under WEP in 3 "growth centres".
5. 4 "urban fuelwood and pole plantations". Total 11,000 ha. Begun in 1980 under WEP inside Forest Reserves for Blantyre, Zomba, Lilongwe & Dowa (for Lilongwe).

ZAMBIA

1. "Lusaka Woodfuel Project". A proposal for initially 7,500 ha, later up to 20,000 ha within Kamaila, Chisamba, Karubwe and Chipilepile Forest Reserves north of Lusaka. Eucalyptus and hybrids. Project never received external funding and did not begin.
2. "Zambia Fuelwood Plantation Project". A proposal incorporating aspects of the Lusaka Woodfuel Project and proposals for woodlots in other towns. Proposed for the Fourth National Development Plan.

TANZANIA
Dar es Salaam
Ruvu North Forest Project. Started 1965, all destroyed 1973–4 due to drought and termites. Started again 1981 with SIDA funds. Now 200 ha (eucalyptus *terecticornis* and *cassia sianiea*).
Dodoma
CDA green belt. 7,000 ha so far; 22,000 ha planned. Enrichment planting of natural woodland using local and exotic species.

Iringa
Municipal tree plantation. About 100 ha(?), started 1983. Eucalyptus.
Mbeya
Town council woodfuel plantation dating from 1950s or 1960s(?). Coniferous softwoods.

Costs of Plantation Establishment
Malawi (1984 figures)

Cost of establishment in first year	– MK 500 per ha
Post-establishment costs years 2–5	– MK 200 per ha pa
Post-establishment costs years 5–8	– MK 150 per ha pa

Zimbabwe (1986 figures)

Establishment costs (plus overheads)	Z$ 1,072 per ha
Post-establishment costs – annual maintenance	Z$ 40 per ha
Post-establishment costs – third-year weeding	Z$ 60 per ha
Post-establishment costs – harvest and coppice	Z$ 246 per ha

Yet peri-urban plantations face a number of serious problems. As single stands controlled by the government or a local authority, they are competing with other producers in areas close to urban markets and often displace other productive activities in the area such as livestock and agriculture. People displaced by the plantations may at times be integrated into the forestry labour force but, more generally, they join the mass of urban dwellers, thus increasing urban fuelwood demand. Plantations have often occupied land more suitable for agricultural production and on more than one occasion they have been sited on land not suitable for the tree species chosen. Great care has to be exercised in the choice of sites and the species to be planted.

The crucial arguments against peri-urban plantations refer to a comparison of costs when measured against other supply-enhancement options. A detailed exercise of this kind was under-taken for Tanzania.[7] A summary of the findings are presented in Table 8.4, although it needs to be stressed that these are pre-project exercises and we still await the proof of the findings in practice. However, the results on paper are quite staggering. Two alternatives were considered: smallholder woodlots using a *taungy* system whereby food and tree crops are produced in forest land; and the management of natural forest involving enrichment planting. These both produce a break-even stumpage price at under a third of the cost of the peri-urban plantation. Furthermore, the smallholder model includes the costs of food production and harvesting but only counts the wood production as benefit. Yet the gap between the cost of producing fuelwood compared with "mining" (i.e. depleting the resource without replenishing it) is still huge. Where paid at all, the stumpage charge for collecting wood from natural forests is only ten Tanzanian shillings per cubic metre, which represents less than a tenth of the cost of producing it either by smallholder woodlots or by improved natural-forest management with enrichment planting.

Table 8.4 Economic Comparison of Fuelwood Production Methods for Urban Markets in Tanzania

(Data per hectare)	*Peri-urban plantation*	*Smallholder woodlots*	*Forest management and enrichment planting*
Seedlings	2,500	2,000	–
Rotation period (yrs)	26	26	20
Wood produced in rotation (m³)	395	395	61[a]
Average annual wood produced (m³/yr)	15.2	15.2	2[a]
Establishment costs (Tsh.)	24,934	6,695	1,359
(of which for mechanization)	(44%)	(3%)	(4%)
Discounted costs (Tsh.)	31,036	6,336	2,499
Discounted production (m³)	99	67	27
Discount rate used	(10%)	(15%)	(10%)
Break even stumpage price (Tsh./m³)	315	95	93

[a] 21 m³ is produced in Year 1 from thinning and firebreak clearance; production thereafter is 40 m³ for the 20 year rotation.

Discounted costs and production are for the full rotation.

SOURCE: UNDP/World Bank, *Tanzania: Urban Woodfuels Supply Study, Vol. I*, Washington DC: World Bank Household Energy Unit, 1987.

Improved management of natural woodland appears to offer a more fruitful route for enhancing urban fuelwood supply. It is cheaper, of proven ecological viability, and provides a range of benefits from nature conservation to tourist potential. Enrichment planting has already been mentioned as one of the techniques that can be employed and we saw a clear case of how this could work earlier, along with other improved techniques, in the example of the Dedza and Chongoni Forest Reserve in Malawi. Other management methods to consider are: improved extraction techniques (in particular, selective rather than clear felling); replanting of especially vulnerable areas; the rotation of areas of extraction to permit natural regeneration; environmental conservation techniques to protect soil quality and to preserve the water cycle; and where charcoal is an important urban fuel, improved kilns to produce the same amount of charcoal from less wood.

Apart from these technical considerations for the improved management of natural woodland, the legal and social aspects are also of concern. Government legislation thus far has concentrated on restricting access to natural woodland rather than opening up areas to local communities. This could involve placing natural woodland under the jurisdiction of sections of the community who will be responsible for managing the asset and reaping the benefits. To be effective, however, changes of tenure would have to be introduced, extension and advice would need to be provided to develop management skills, and some form of licensing of wood producers would need to be enforced by the local community. Such measures are already being proposed elsewhere in Africa, for example in Niger.

One-off mining of otherwise protected forest areas is a bridging solution which can also be considered as a stop-gap measure in certain circumstances. Mining involves the exploitation of a wood resource without replanting and at a rate which exceeds natural replenishment. Planned one-off mining can be distinguished from the unplanned mining of fuelwood resources which creates such damage to the environment.

In some countries, remote areas of dense forest can be used if transport can be given access. Tapping these resources can bring relief to smaller areas of "besieged" natural woodland near to existing urban settlements which are under intense pressure in terms of energy demands. It can buy time. Basically, much greater attention needs to be paid to improving the management of areas currently supplying fuelwood to the cities.

Another possibility for enhancing supply options is encouraging smallholder-farmer fuelwood production. Generally speaking however, small-scale production of fuelwood for the market cannot compete economically with alternative land-use options for areas near urban centres. Market gardening, dairy-cattle, pig raising and other high-value food production will normally be far more profitable. Only with the growing of construction poles can tree production for the market compete in certain locations. Yet the production of construction poles around the urban periphery automatically produces a fuelwood off-take for the families concerned, relieving the pressures of domestic energy demand upon the immediate urban hinterland.

The evidence of the limited potential for commercial fuelwood production from small farms has been provided from studies in Malawi. An analysis was made of the potential returns to smallholder-farmers of fuelwood and alternative production possibilities from their land. There is little evidence of the viability of commercializing fuelwood production from a smallholder base. To grow enough trees to meet the needs of household consumption in Malawi would require 500 trees to be planted on 0·2 hectares of the limited land generally available to smallholder farmers. Under the severe land pressures in Malawi, this would mean foregoing earnings from maize production upon the land allocated to trees. Given the cost of fuelwood in the cities, a relatively high net loss would occur to family earnings by switching from maize to fuelwood. The greater use of agroforestry techniques on small farms will, however, contribute indirectly to the urban fuelwood supply by relieving pressure on the woody-biomass resource base.

Ways can be found of better utilizing wood as a residue from other production activities, in particular from land clearance for agricultural production or construction purposes. In Swaziland, for example, when land was cleared for the huge sugar estates in the low veld, the wood was simply burnt. On the large farms, such as the tobacco estates in Malawi which use an enormous quantity of fuelwood for curing, measures can be taken to oblige them to grow their own fuelwood plantations, thus reducing the pressure on the indigenous tree resources from which the city's fires are made. In certain areas, far better use could be made of the wood residues from commercial forestry operations. The economics of this will need to be carefully explored, in particular aspects of transport and marketing which remain the crucial bottleneck for improved urban fuelwood utilization. In Zimbabwe, Angola, Swaziland and Mozambique, commercial timber companies currently use the waste in their production processes and the workers receive wood for domestic use as a perk to the job.

There are, then, various ways of enhancing supply of fuelwood to urban centres. As the indigenous forest resources grow more and more depleted, the economic viability of peri-urban fuelwood plantations may need to be reassessed. Supply enhancement and fuelswitching appear to offer the best, but by no means the only, policy options for tackling the problem of urban fuelwood demand and relieving pressure upon the woody-biomass resource. In the same way that agricultural residues provide an energy safety-net for people in the countryside, urban life produces its own combustible residues: old newspapers, packing cases, tyres, cardboard boxes, packaging of every kind. These are all available for incineration in the hearth and all are used, especially by the poor. Whether there is scope to improve the availability of such combustible waste as a matter of policy, we simply do not know.

One final source which combines both supply enhancement and fuelswitching, is the use of various forms of agricultural residue. Figure 8.1 illustrates possible routes for the treatment and conversion of agricultural residues for energy production. These cover a broad spectrum of biomass materials produced as by-products from agriculture including cereal stalks and cobs from cereals, coconut shells and husks from woody crops, groundnut straw from green crops, groundnut shells from crop processing, and finally animal dung. In the rural areas, agricultural residues provide an energy safety-net, a fuel of last resort. Only in virtually treeless Lesotho does dung play a significant role in rural and even in peri-urban domestic energy-consumption. Interventions to promote the greater use of agricultural residues for rural domestic energy consumption are fraught with difficulties, essentially because they are also needed as fodder and fertilizer. Such interventions are generally unworkable as a policy objective.[8] However, greater industrial use of certain agricultural residues is certainly viable, particularly the material produced from agro-processing.

We do not know very much about the patterns of use of residues for industrial energy in the SADCC region. Table 8.5 provides a list of some of the more important ones. The range of available technologies is wide and can be divided roughly into two categories:

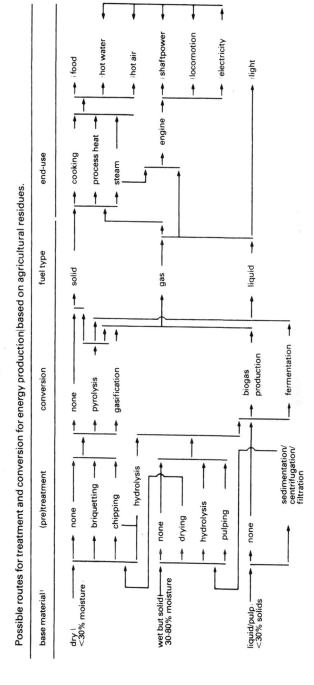

Figure 8.1 Possible routes for treatment and conversion for energy production based on agricultural residues.

Table 8.5 Use of Agricultural Residues in Industry

Sugar Processing:	Bagasse is the main source of fuel in sugar processing. Large amounts of energy are needed for boiling and evaporating cane juice. Mechanical energy is also required to run crushers and other equipment. In efficient plants, the energy from burning the bagasse is enough to run the entire mill. Most SADCC countries grow at least some sugar, the largest producers being Swaziland and Zimbabwe.
Oilpalm Processing:	Oilpalm processing requires heat for sterilizing the fresh fruit bunches, digesting the fruit and drying the nuts and kernels. The fibre and kernel-shell by-products, however, are more than enough fuel to provide all the energy needed, and in some cases surplus energy is generated. In the SADCC region, the oil-palm industry is mainly restricted to Angola.
Coconut Processing:	The shell and husk from coconuts are used to dry the copra. When efficient driers are used, and especially when sun drying is employed as a preliminary step, there is a considerable surplus of coconut residues available. The main coconut producer in the SADCC region is Tanzania.
Cottonseed Processing:	In Zimbabwe and Malawi, cottonseed husk is used as a boiler fuel and also sold for use in other nearby factories.

conversion technologies where the residue is transformed into a different fuel source; and combustion technologies involving its direct burning.

The three most common conversion technology options are biogas, gasification and briquetting. Producing biogas involves the fermentation of organic materials in an oxygen-free environment.[9] The mixture of methane and carbon-dioxide gas that is produced can be used for a variety of things such as cooking, lighting, running diesel and gasoline engines. These biogas units make a significant contribution to energy use in China and India but thus far remain experimental in Africa. Tanzania has the greatest experience of biogas digesters in Africa, with several hundred already installed. Gasification is a thermal not a biological process. It still remains an experimental technology within the SADCC region and is normally employed for driving engines. A prerequisite to developing widespread use is encouraging local production capability. Briquetting compresses combustible materials into a denser and more compact form, making it easier to transport and store, and provides a more combustible material. Trials involving sawdust briquetting have taken place in Malawi and Mozambique, and there are plans for coffee and rice-husk briquetting in Malawi. In Angola, a sawdust-briquetting project is under study and there is a proposal to test different woods for sawdust briquetting in Zambia's Fourth National Development Plan. There is little practical experience of this form of energy use within the region.

Combustion technologies have been used in the SADCC region mainly for industrial energy use. In particular, the vast plantations in Swaziland that produce wood pulp, have mechanisms for burning the fuelwood waste. Also, Swaziland's sugar mills, along with those in Zimbabwe, make extensive use of bagasse. In sum, agricultural residues for industrial use offer fuelswitching options which can employ the biomass residues from agricultural production more effectively. They offer another option for supply enhancement. However, while they can make an impact in specific locations, they are unlikely to make a significant contribution to urban energy use in anything other than the short term.

FUEL CONSERVATION

Energy conservation offers a number of possibilities for reducing the growth of urban fuelwood consumption. The major conservation measures that we will consider are improved domestic stoves using firewood, charcoal and kerosene, and improved kilns for charcoal production. Energy-conservation technologies can also be used to encourage a measure of fuel substitution. Improved-stove programmes can provide a level of protection for consumers against rising prices.

There are many lessons to be learnt from the experiences gained from improved-stove projects within the SADCC region. Indeed, every single country has tried such a project on an experimental basis. Yet the emphasis has generally been upon improving energy efficiency for stoves in rural areas instead of urban stove programmes. These are the five main reasons why the idea of rural improved-stove programmes has not take off.

1. A free alternative exists in the form of the traditional three-stone fire and cash to purchase a new stove is a scarce resource in the countryside.

2. Rather than incur an unnecessary expenditure by purchasing an improved stove if fuelwood scarcity threatens, traditional fire-management practices will be improved within the household. This means that stove programmes which concentrate on rural households face the generally insurmountable problems that both the fuel itself and the stove are normally non-monetary items of expenditure.

3. Improving cooking efficiency with new stove designs generally means constraining the flame to intensify the heat output. This inevitably interferes with the multiple functions that traditional fires perform, in particular heating, lighting, and a family, indeed a community, focus in the evenings. There are additional benefits with the open fire such as the smoke reducing the problem of insects in thatched roofs.

4. Most of the projects in the region have concentrated upon research and development, and not on mass production and dissemination.

5. The three-stone fire is portable, amenable to complex fire-management practices and it burns all forms and shapes of fuel, something that cannot be said of all improved stoves.

Of course, there is not necessarily a clear distinction between stove designs for use in rural and in urban areas. Generally however, insufficient attention has been given to specifically targetted urban-stove programmes for which the chances of success would appear to be far greater. First there is a more generalized commodity environment with more cash around. Secondly, there are often more constraints upon the use of open fires, given the greater density of population and the more enclosed urban environment.

Encouraging the use of efficient stoves is intended to achieve more than merely economizing upon fuel. It can help reduce the labour-time needed to procure fuel; produce a healthier environment in which fuel is prepared and cooked; and the programmes also generate employment. When stove programmes are being developed, it is important to have a clear idea of the particular set of objectives which are to be met.

It is essential now to switch the emphasis from research and development to mass dissemination. Here there is much to be learnt from the successful Kenyan experience. In the early 1980s, the traditional jiko stove was improved by a clay insert. A number of models were built and tested. A debate then emerged about whether these should be mass-produced by modern industry, which would have a negative impact upon the large number of informal-sector jiko-makers in Nairobi, or to integrate the new design within informal-sector production. The outcome was a successful compromise with mass-production providing the clay inserts for the informal-sector jiko-makers. A number of benefits followed, with the clay inserts extending the hearth life of the jiko and also ensuring that the jiko-makers conformed to certain standards. The centralization of a part of the production process together with the utilization of the vibrant informal-sector producers for the component, the availability of the product through both formal and informal marketing channels, and the public debate that the issue generated, all contributed to a high market profile and a successful programme. Production soon exceeded 20,000 stoves per year. The success of the programme was primarily based upon targeting the urban market and discovering a particular market niche, where both fuels and stoves were already a commodity. The large-scale production of the improved urban stove is now seeping into the surrounding rural areas.

Another change of emphasis required within existing stove programmes is a commitment to exchange information and experiences between the different countries. The SADCC offers just such a platform to promote and benefit from shared experiences. This will need to be continually built upon as the exchange of information has not always been as effective as it might.

Improved-stove production offers a great opportunity for expanding formal and informal employment in the urban sector. Given the existence of a number of satisfactory designs, the problem is that of minimizing production costs so that the potential consumer can see the benefits of his or her investment in fairly immediate terms. For the informal-sector producers to flourish, their special needs must be taken into account. They are often heavily reliant upon scrap metal and this has to be made available in sufficient quantity and quality.

This is a potentially important growth area for the informal sector which as a whole is going to receive increasing support from the major international agencies into the 1990s.[10] Of particular importance will be the intermediary sector at the interface between the so-called formal and informal sectors. The Kenyan stove programme is a classic interface. There are many opportunities for improved stoves in urban areas and not simply for domestic use. The possibilities are enormous, ranging from roadside fast-food vendors for the shanty-town commuters, to bakeries, restaurants and hotels. Encouraging energy conservation through the dissemination of improved stoves for urban use is a comparitively cheap way to encourage economic development and a more rational urban growth which reduces the pressure upon the rural environment by decreasing fuelwood and charcoal demand.

An additional means of reducing the pressures on the rural woody-biomass resource is improving the efficiency of kilns for charcoal production. Of course, increased reliance upon charcoal may have both positive and negative effects. One recent study has suggested that the saving in transport costs and increased energy efficiency which comes from transforming fuelwood to charcoal, mean that the charcoal supply area is potentially 100 times greater than that of firewood.[13] The converse of this argument is that charcoal production has a larger negative environmental impact on the urban hinterland than fuelwood and the increased use of timber that it requires accelerates the process of deforestation.

IMPROVING MARKETING AND TRANSPORT

Fuelwood and charcoal markets are highly complex and understanding the chain from producer to consumer is a difficult affair. Each stage involves its own mix of capital resources, levels of skill, labour force composition, and dominant concerns. Obtaining information from the participants is complicated by the fact that many of their activities are on the margins of legality. The fuelwood trade is essentially an informal-sector activity and its entrepreneurs do not keep records or accounts. Many of those involved drift into the trade for only short periods of time and some may be illegal immigrants. The workforce in the fuelwood trade is in a state of constant flux and the evidence appears to show that the work is extremely labour intensive. This seems to indicate that as a source of income it is a last resort. Therefore, the possibilities of reducing labour costs in the chain are not great.

To date, the record of government interventions in fuelwood markets in order to control prices or sources of supply has not been one of great success. One such scheme was Angola's attempts to license fuelwood traders. Licensing has the effect of reducing the number of traders in the market, facilitating monopolization and price-fixing. At the same time, it may also promote an "informal" marketing system. Generally, too little is known about the transport and marketing structure and there always appear to be ways of avoiding controls. To intervene effectively, the costs at every stage and the extent of the mark-up have to be known. In Luanda in 1986, for example, charcoal was sold by producers at the roadside for 125 kwanzas per kilogram and transport costs were up to 40 kwanzas. The urban retailers paid 250–375 kwanzas but sold it to customers for 800–1,200 kwanzas per kilogram.

An extended chain exists between the producers and consumers. This has a number of distinct stages involving different forms of transport and a number of breaks in the bulk. At every stage, a number of costs is incurred and the price to the consumer increases. In order to intervene effectively, with the aim of improving fuelwood supply to the cities, the process has to be properly understood. Land clearance and wood cutting are the first step. Trunks and branches are then split or the wood is gathered together in clamp sites for charcoal production. Those products have then to be transported to a collection point. From there, it is transported to the cities and moves from wholesale dealers to sub-wholesale and retail traders. At every stage equipment, machinery and labourers all add to the cost. Finally, logs are split down to small bundles of firewood and sacks of charcoal into small piles of no more than a handful – enough to cook a single meal.

Many people are employed along this chain and any outside interventions to improve the security of supply or to cheapen the end-product must be wary of creating unemployment. Yet in a number of areas throughout Africa, traders and transporters occupy monopoly positions and can greatly inflate the cost of fuelwood and charcoal to the consumer. Traders generally purchase directly from the producers, either from charcoal-makers or woodcutters. Prices are negotiated and vary according to the season.

Production is inevitably more limited in the rainy season. The reasons are complex but strike at the heart of rural life. Many of the charcoal producers and woodcutters are part-time workers, relying for much of their livelihood upon subsistence agricultural production. In the rainy season they are busy with their farming. Working conditions in the wood sector are difficult at this time as the wood is wetter, takes longer to carbonize and gives a lower yield. Naturally, transportation in the rainy season is also far more difficult. Seasonal variations are not simply a reflection of relative costs. In the cold season, the higher demand for fuelwood for space heating encourages the merchants to increase prices. In a number of SADCC cities, the highest seasonal price is 10–20 per cent above the "normal" price. But who fundamentally determines the price settings? Producers and marketers claim that it is the transporters who work together to ensure higher returns. Costs can be reduced if "back-hauling" can be guaranteed, i.e. that the trucks do not travel one way without a cargo.

Often, traders have a monopoly over transport and market access. There appears to be little difference between the prices paid to different producers at any given site yet between sites there are likely to be definite price differences. These are determined by the distance that the fuel has to be transported and the ease of access and movement. In the case of the latter two factors, distance off the road and the season in which the collection occurs will alter the price paid. Interestingly, the price is always determined by volume and not by weight. This is a feature of the transition period as in a modern economy standardization and commoditization produce sale by weight not volume.

The price in the urban marketplace provides the ceiling under which prices paid along the chain are determined. To illustrate this, let us look at the charcoal market in Lusaka. The prices given are those current at the time the survey was conducted in September, 1986. Charcoal coming from Kamaila FR, 40 kilometres from Lusaka with about 7 kilometres of the journey off the road, fetched a price of 5 Zambian kwatcha per bag. At Chongwa, 65 kilometres from the market including 20 kilometres off the road, the price ranged from 4·5 to 5 kwatcha per bag. By the time one reached Mumbwa, 100 kilometres away with 8 kilometres off the road, the price paid to the producers was down to 3·50–3·70 kwatcha per bag. In other words, there is a trade-off between the producer's margins and the trader's margins depending upon the transportation costs. Another feature of this chain is that the volume criterion is a flexible measure for squeezed margins. In Harare in Zimbabwe for example, the size of a *cord* of wood (usually estimated at 128 cubic feet) has been diminished, according to local retailers. The actual transportation of fuelwood adds considerably to costs but offers few avenues for intervention. Inadequate transport facilities are a general feature of the region and improvements in this area are part of the wider development effort. Yet if ways can be found of shortening the chain then the price paid to the producer can be increased.

In some cities, the location of the various markets will reflect the different transport and producer costs. Hence, in the peri-urban area of Mbezi (15 kilometres from the the city centre), people pay higher prices than in the central Dar es Salaam charcoal markets. Economies of large-scale transportation and marketing permit a lower price in the central market. Out in Mbezi, people paid 240–260 Tanzanian shillings for a bag of charcoal, rather than 170–180 shillings in the city centre markets (in September, 1986). This is because they rely upon people bringing in the charcoal on bicycles or they pay premiums to lorries passing through on other business for temporary transport assistance.

Frequently, the fuelwood marketing chain is short-circuited by an entrepreneur who may see the marginal advantage in purchasing wholesale from the producers and selling directly to small retailers from the back of a lorry, by-passing the official markets and big retailers. Within his own economic rationale, he has to hire a lorry to transport the fuelwood or charcoal from the site of production to the market. He hires the lorry for one or two days and so may consider using the mobility that this provides to sell the wood directly to the small retailers. Interventions to support this trend would have to bring more traders into the business to prevent a consolidation of transporters' monopolies. In addition, there are small traders emerging who illegally cut wood from municipal land, impinging upon peri-urban plantations and urban amenities.

So what scope is there for interventions? Essentially ignoring or trying to replace market systems will be ineffective. Interventions must work within the existing system to increase competition, improve efficiency and attempt to by-pass certain stages in the chain from producer to consumer. On the latter point the greatest opportunity exists in closing the gap between wholesalers and retailers, partly because the retailers' marketing margin with the final break of bulk is the highest in the sequence. Yet they have to carry the cost of losses, such as those incurred by the break of the bulk of a sack of charcoal which produces "fines" (dust) which cannot be sold. Consumers' co-operatives may be of help here to the urban poor who frequently only have the cash available to purchase enough fuel for the day's cooking, which imposes a very high cost to those least able to afford it. Where consumers' co-operatives have sufficient resources, the cost of wood will be minimized if they are able to purchase directly from the producer and provide the necessary transportation.

CONCLUSION

To achieve the objectives of ensuring a continuing supply of fuelwood to the growing urban centres while limiting the negative effects upon the rural areas, requires a package of interventions. The two most important arenas for action are fuelswitching and supply enhancement.

Fuelswitching will depend upon ensuring the availability and reliability of supplies of different fuels and their appliances at affordable prices. In so far as there is a switch for some away from fuelwood, this will help reduce the pressure upon tree resources in rural areas. Yet fuelwood will remain a significant urban domestic fuel well into the twenty-first century.

Enhancing supply availability is therefore also of great importance. Crucial efforts should be focused upon improved management of the existing supply areas, in particular of natural woodland. Better ways need to be found for utilizing fuelwood as a residue from other production activities such as land clearance, or from forestry plantations.

Of lesser importance are conservation methods, which for improved urban stoves could go hand-in-hand with fuelswitching strategies. There are some opportunities for improved charcoal kilns but problems exist in technical efficiency and the potential social impact within the workforce as new technologies frequently benefit the already advantaged. Interventions in transport and marketing will require careful prior preparation and should concentrate upon using existing market mechanisms.

REFERENCES

1. G. Leach and R. Mearns, *Bioenergy Issues and Options for Africa* (London: IIED, 1988), p. 183.
2. Ibid., see Chapter 4.
3. D. Barnes, *Understanding Fuelwood Prices in Developing Nations* (Washington DC: World Bank, Agriculture and Rural Development Department, unpublished, 1986).
4. G. Leach and R. Mearns, op. cit., p. 146.
5. UNDP/World Bank, *Tanzanian Urban Woodfuel Supply Study*, Vol. I (Washington DC: World Bank Household Energy Unit, 1987).
6. T. Byer, *Review of Household Energy Issues in Africa* (Washington DC: World Bank, unpublished, 1987). Note that the figures in brackets after the countries refer to the year on which the calculation was based.
7. UNDP/World Bank, *Tanzanian Urban Woodfuel Supply Study* (Washington DC: World Bank Household Energy Unit, 1987). The findings of this study are ably summarized in Leach and Mearns, op. cit.
8. For general background reading on agricultural residue energy use, see; G. Barnard and L. Kristoferson, *Agricultural Residues as Fuel in the Third World*, Technical Report, No. 4 (London: Earthscan, 1985); J.F. Henry, A. Talib and K. Ford, *Handbook of Biomass Conversion Technologies for Developing Countries* (UNIDO, 1984); D. Hughart, *Prospects for Traditional and Non-conventional Energy Sources in Developing Countries* (Staff Working paper, No. 346 (Washington DC: World Bank, 1979); J. van der Ham, J.B.P. de Loor, M. Flack, *Energy from Tropical Crops*, Tropical Crops Communication, No. 8 (Wageningen: University of Agriculture, 1985).
9. On biogas usage, see; Escape, *Guidebook on Biogas Development*, Energy resources development series, No. 21 (New York: United Nations, 1980); G. Foley and G. Barnard, *Biomass Gasification in Developing Countries*, Technical Report, No. 1 (London: Earthscan, 1983).
10. S. Please, "Can sub-Saharan Africa achieve sustainable growth with equity?", World Bank seminar for UK academics, University of Manchester, 22 March 1988.
11. G. Leach and M. Gowen, *Household Energy Handbook* (Washington DC: World Bank, 1987), p. 107.

9. Conclusions

A NEW APPROACH IS NEEDED

Fuelwood is the people's source of energy. Most people in the SADCC region depend upon it for survival. It provides four-fifths of all the energy consumed. It is mainly used in the household for transforming the fruits of production for human consumption and for the provision of heat and light, and it will continue to be so for years to come. Yet, as more and more land is cleared for agriculture and settlement, what was once a common and free resource is now under stress in a growing number of places within the region.

The problem is not general, it is specific to people and to place. Effective solutions cannot, therefore, be general either, they, too, must be specific to people and to place. The costs of woodfuel scarcity have for too long remained hidden. Deforestation has increased the time that women (but also children) must labour to collect it. As more time is spent on collection and on fire management, less is available for work in agricultural production, child rearing, housekeeping and study.

Deforestation has an impact on many different aspects of people's lives. Trees provide many benefits and have many uses and woodfuel is only one of them. Deforestation brings with it a deadly cost in environmental degradation. Trees are vital in guaranteeing the earth's fertility, remove them and soil erosion takes its toll. This includes many things from declining food production to the increased expenditure of foreign exchange on, fertilizer and pesticides.

What all this means is that shortages of fuelwood are a symptom of a much broader problem. They cannot, therefore, be tackled, as has been tried in the past, merely by growing plantations to enhance the supply of woodfuel or by developing cookstove programmes to conserve existing supplies. Useful as they may be, these are narrow responses within the broad spectrum of woodfuel policy interventions.

Energy plantations are expensive and cannot usually compete either with the supplies from natural woodland or with the various forms of woody biomass in and around the farm, both of which are free of production costs. Solutions to the problem of shortages which are based solely on producing fuelwood as a commodity are bound to be ineffective so long as free supplies remain. Similarly, improved stoves are unlikely to have a big impact in rural areas where the three-stone fire, or its local equivalent, can be had free.

The fuelwood trap awaits the unwary. As soon as the fuelwood problem was defined as the "other" energy crisis facing much of Africa (in particular areas of east, central, southern Africa, the Horn and the Sahel), governments and aid agencies rushed to find fuelwood solutions. We have tried to show that many of the so-called solutions could not achieve the desired results. Instead we have urged that a new way of understanding the problem be adopted. This has to begin by properly defining the role that trees and woody biomass play within rural-production systems. To do this requires an understanding of people's needs from trees. A greater willingness by people to grow more trees and hedges will follow an attempt to meet those needs. An indirect benefit of such an approach is that fuelwood will be inevitably provided in the process.

The problem is much wider than a narrowly defined energy issue, however. It is about how to produce sustainable development. Hence, a proper understanding of the fuelwood puzzle means finding solutions to a wider array of development problems. Such an approach will contribute to a genuinely integrated development effort. Lest this approach be heralded too soon as yet another miracle solution, only to bring disappointment in years to come, certain cautionary notes must be sounded. First, sustainable development, as Michael Redclift has recently pointed out, is not an unproblematic concept, as it "draws on two frequently opposed intellectual traditions: one concerned with the limits which nature presents to human beings, the other with the potential for human material development which is locked up in nature."[1]

Achieving economic development in harmony with the natural environment is no easy task. This is particularly difficult whilst donors, aid agencies and governments apply rigorous cost-benefit analysis for funding projects or programmes whose real benefits may be long-term and diffuse.

Secondly, the new approach outlined in this volume is just that, an approach. There are signs of success where this has been applied in practice but it is still early days and much more on-the-ground experience is required before any final judgement can be passed. As Lloyd Timberlake has commented, "there are promising agroforestry projects in Africa, the success or failure of which can only be judged after about another decade."[2] While there appears to be a fairly coherent set of strategies to pursue in tackling the problem in the rural areas, the issue is far less clearcut in the urban locations. The data base for urban areas is weak and much more basic research is needed on urban energy-consumption.

It is one thing to have a clear idea about the strategy to pursue but quite another to be able to put it into practice. What are the absolutely essential ingredients needed to ensure success? There appear to be three: agents on the ground who are well-versed in this new approach; a genuine commitment towards ensuring popular participation; and integrating the efforts of the separate institutions.

Much will depend upon the training and motivation of agents for change. Lea and Chaudhri in their extensive study of many rural development projects concluded that it is generally an outside institution or individual that will act as the catalytic force.[3] Without a body of individuals able and willing to launch such initiatives, change will never occur at the necessary rate. Yet the outsider must have the right attitude and approach. As Paul Harrison has observed,

> Those who succeed in Africa also share a common attitude to Africa's people. They have respect for ordinary African peasants – not only theoretical respect for their human dignity, but practical respect for their views and their wishes, and respect for their accumulated wisdom and traditional practices.[4]

The second important ingredient for success is participation, beginning from the word "go" and lasting right through every stage of the process. Bryant and White have summarized a checklist of lessons derived from a survey of experiences in ensuring participatory rural development that benefits small farmers.[5]

- Improve listening and communication skills.
- Build on the natural interests and priority goals of peasants.
- Find ways to ensure participation is seen as a benefit in itself, and not purely a cost.

- Design projects so they are small and simple enough for people to actually work together.
- Work through local organizations, preferably existing ones, and if these do not exist then build them.
- Assign or train staff to facilitate community development.
- Gain (voluntary) resource commitments from local groups.
- Build coalitions with potential political supporters in the community and with clients and other beneficiaries.
- After analysing the situation surrounding a project or programme, design the most appropriate organizational means for delivering it.
- Design ways to protect or buffer local projects from local elites who would otherwise co-opt its benefits.

One of the major current difficulties which can hamper governments from putting such an approach into practice is the gap in institutional knowledge. To parody the situation: forestry departments do not *like* people and energy departments do not *know* people. Yet, ironically, it is these two departments which are doing most to understand and tackle the fuelwood problem. The department that is best equipped to provide the level of practical intervention required on the ground, namely agriculture, does not *like* or *know* trees! A far more integrated institutional approach is required across many sectors including agriculture, environment, land and natural resources, rural development, energy and forestry. Urban planning and industry must obviously be involved when dealing with the urban energy situation. Better co-ordination could provide a strong lead from the top but the real activity has to take place at district and local level. Decentralizing information and expertise is vital to success and the only way to ensure both leadership and dynamism at all levels of popular participation. The way to begin is by having faith in the local farmers.

DEVELOP SEPARATE POLICIES FOR RURAL AND FOR URBAN AREAS

Separate fuelwood energy policies are required for rural and for urban areas. This is because in rural areas fuelwood is both produced and consumed with few viable alternative domestic energy supplies. In urban areas, fuelwood is only consumed and alternative fuels and the cash to buy them are more readily available.

Rural areas

Policy interventions in rural areas depend upon understanding the role of woody biomass within integrated production systems. Most fuelwood comes from farmland and not from gazetted forest or from plantation areas, but it is only one of many things that rural people need and is rarely their primary concern. But people can be encouraged to grow more trees where it is possible to show that by doing so these primary concerns can be addressed, whether in terms of cash earnings, increased soil fertility through intercropping, the production of construction poles or of fodder for their cattle. This is more cost effective, it produces fuelwood nearer the site of consumption thus diminishing the labour of collection and it can slow down the process of environmental degradation. In sum, the fuelwood problem is best tackled indirectly by improving land-use management practices.

Such an approach is based on listening to people define their own priorities and on intervening in ways which try to meet their needs. Beyond this, however, it has to take into account the responses of smallholding farmers to the problem of diminishing supplies of woody biomass for they have developed their own management practices. Paul Richards has demonstrated the emergence of innovative practices in indigenous agriculture[6] and our own research has confirmed that they are to be found in the management of woody biomass as well. The picture is not, however, uniform. Generally the highest levels of tree management are to be found in areas of high-intensity arable production and lower levels in extensive grazing land-use management systems.

The route to effective action lies in identifying the land-use and woody-biomass management systems correctly and in appraising the constraints on and the potential for improved productivity in those systems. But it depends on the closest possible involvement of local people and in a sharing of knowledge.

Urban areas

Urbanization is increasing rapidly throughout the region. Within the next twenty years the urban populations of a number of SADCC countries will exceed their rural populations. Fuelwood will continue to be a major source of energy for the towns and cities.

In order to deal with the problem of energy hardship among the urban poor policies are required which are designed to lower energy costs, to increase the security of fuel supplies and to reduce the impact of urban fuelwood demand on rural resources and people. Policy interventions are possible in four key areas. In order of importance they are: fuelswitching, supply enhancement, conservation and improvements in transport and marketing.

A variety of fuels can be substituted for wood in cooking including kerosene, bottled gas, coal and electricity but the cost of these is not the only basis for consumer choice. Many people do not have the money to buy an expensive energy device and the insecurity of fuel supply is often a major concern. A policy of fuelswitch usually works best if it follows existing trends. The overall goal is to protect the interests of consumers by increasing both the range of fuels available and the security of their supply.

The best tools for this policy are the development of fuel distribution systems and of the supplies of affordable appliances. Such development should also be seen as part of a programme of job creation. In the longer term it is important to develop local fuel resources and greater cooperation over commercial energy supplies throughout the region. An active policy of fuelswitch in urban areas would seem to be promising but it needs careful preparation, particularly in terms of the comparative foreign-exchange costs of the different options.

A policy for enhancing supplies should aim at improving the management of existing areas which supply fuelwood to the cities. Better means can be sought for making use of the woodfuel produced as a residue from other forms of production: for example, land clearance, toppings and cuttings from commercial forests etc. Finally it may prove to be possible to open up new fuelwood sources on plantations and farms.

Conservation will have less of an impact, but governments can develop policies to encourage the dissemination of improved stoves and improved charcoal kilns. These policies would need clear objectives and target groups. Finally improvements are possible in marketing and transportation. Here the aim would be to shorten the chain between supplier and consumer and thus to reduce the mark-up in price.

These then are the general conclusions to be drawn from the study. The challenge now is to begin to put these policies into practice and then to refine them in the light of experience.

REFERENCES

1. M. Redclift, *Sustainable Development. Exploring the Contradictions* (London: Methuen, 1987), p. 199.
2. L. Timberlake, *Africa in Crisis* (2nd edition, London; Earthscan Publications, 1988), p. 194.
3. D. Lea and D. Chaudhri, *Rural Development and the State* (London: Methuen, 1983).
4. P. Harrison, *The Greening of Africa* (London: Paladin, 1987), p. 301.
5. C. Bryant and L. White, *Managing Rural Development: Peasant Participation in Rural Development* (W. Hartford CT: Kumarian Press, 1984).
6. P. Richards, *Indigenous Agricultural Revolution* (London: Hutchinson, 1985).

Appendices

Appendix I
Agroforestry Options For The SADCC Region

This appendix will outline some of the technical agroforestry options which can be employed to improve the management of woody biomass within the region.

NEW SPECIES

Looking at traditional tree-growing practices, it is noticeable that in many areas the range of species planted is quite limited, eucalyptus and black wattle being the two most common. Clearly there is major scope for enriching local tree-growing practices by introducing new species.

Matching the species to local conditions is obviously important if this strategy is to succeed. Demonstrations will also be needed if farmers are unfamiliar with them and are suspicious of their growth potential and properties. Once new species have been proven, however, in many cases they are enthusiastically adopted.

The SADCC region is blessed with a wide variety of indigenous tree species which have a number of intrinsic advantages: they are already adapted and proven under local conditions; people are familiar with their properties and produce; they help to preserve the diversity of the local flora and fauna; they are unlikely to spread and become a weed (this can be a problem with some fast-growing exotic species).

Their major disadvantage is their generally slower growth rates in comparison to some exotics. This means that farmers rarely plant some of the best known indigenous species, such as *Julbernardia paniculata*, *Brachystegia longifolia* and *Soberlinia angolensis*, even though they are recognized as providing a number of valuable and high-quality products.

Not all indigenous species are slow growing, however. The problem is that few have been properly researched, and even fewer subjected to the rigorous selection and breeding programmes needed to identify the best varieties. They are a major resource, which is only now beginning to be tapped.

Excluding exotic species from a tree-planting programme, on principle, is clearly ridiculous. Carefully selected exotic species have a great deal to offer, both in terms of their rapid growth rates and of the useful products they can supply. Eucalyptus is a classic example. Introduced to Africa in early colonial days, it is now one of the most widely planted group of trees. Although it remains the subject of considerable controversy (see Appendix II: Eucalyptus: Friend or Foe?), its merits are clearly appreciated among farmers.

This is not to say that other exotic species could nto do equally well. Indeed, a shift away from Eucalyptus towards a broader selection of species is now widely regarded as a desirable move. There are a number of promising leguminous species which are being actively tested in several SADCC countries. *Leucaena leucocephala, Calliandra caliothus,* and *Gliricidia sepium* are just three examples. In drier areas, the genus *Prosopis* offers good potential and has the advantage of producing large quantities of edible pods.

Of this group, *Leucaena leucocephala* has been introduced most widely to date. It has suffered, however, from being branded a "miracle tree". This has lead to it being planted regardless of environmental or socio-economic conditions, and hence to many disappointments (see Appendix III: Beware The "Miracle Tree"). Clearly, a sensible approach is needed when introducing new species. Promoting a tree which has not been properly tested can do more harm than good, and may undermine the success of a whole programme if the confidence of local people is lost.

NEW TREE CONFIGURATIONS

Another way of improving tree-management practices is to introduce new configurations, in which trees are grown in different locations on the farm and in new spatial arrangements.

Intercropping

Intercropping of trees and field crops is one of the most promising approaches. Provided the right combinations of trees and crops are chosen, it is often possible to achieve higher overall production levels than if the two were grown separately. To many farmers, intercropping is not new. In parts of the Sahel zone, the species *Acacia albida* is commonly grown in among foodcrops. Because of the beneficial effect of the trees, crop yields nearer to the trees can be as much as double those of places further away. In areas of higher rainfall, *Sesbania sesban* has a similar function.

Alley cropping

A new intercropping approach that is currently receiving considerable attention is the technique of "alley cropping". This has been developed by the International Institute for Tropical Agriculture (IITA) in Nigeria. It is more systematic than traditional intercropping methods and is therefore better adapted to agricultural mechanization such as oxen ploughing. Most research has been carried out using the combination of *Leucaena leucocephala* and maize. The production cycle is relatively simple and can be summarized as follows: 1. *Leucaena* is planted in maize fields at the beginning of the rainy season in rows spread 4 metres apart – wide enough to get a plough in between. 2. During the first season the *Leucaena* is suppressed by the faster-growing maize, and so grows only to the stage of a thin sapling. No beneficial effects on the crop can be expected. 3. The maize is harvested, after which the *Leucaena* develops quickly into healthy trees 1–2 metres in height. These shade the ground in between the rows, helping to suppress weeds. 4. At the start of the next rainy season, maize is sown in between the rows of trees. 5. Once the maize has germinated and become established, the trees are heavily lopped, cutting them back to a height of about 30 cm above ground level. Foliage is left on the ground as green manure, while stems and branches are removed to be used as fuelwood or for other purposes. 6. The maize is harvested again, and the cycle repeats (back to stage 3).

Significant increases in maize yields have been noted using this technique, especially on unfertilized fields. A useful quantity of small firewood sticks are also produced. Other benefits include a gradual increase in the organic matter content of the soil, making it more fertile and easier to work, and a reduction in the amount of wind and water erosion. One disadvantage, however, is that it requires increased inputs of labour – something that may be difficult to supply in places where labour is scarce.

Taungya

A variant on this approach is to grow foodcrops within newly established woodlots or plantations. In large-scale forestry plantations this practice is commonly referred to as the "taungya" system, a term originating in Burma where it was first developed. It can also, however, be used by individual farmers.

During the first year or two after a plantation is established, weed competition is often a severe problem, necessitating laborious and expensive weeding. By growing a food crop in between the trees, not only is there a useful additional product but the need for weeding is reduced. Leguminous beans are one of the best crops because they do not compete with the trees for light or soil nitrogen.

In this way, the economics of plantation establishment can be improved significantly. Forest departments have used taungya as a means of reducing plantation establishment costs. In some cases this has been heavily criticized, since the system has often been used in an exploitative way (for example, workers have been provided few benefits and very little security of tenure). But this need not be the case; instead taungya can be used as a system of resource-sharing with the local community. Where taungya is practised by individual farmers, this criticism also need not apply.

Erosion control

Wind and water erosion are a major threat in many parts of the SADCC region. Trees can play an important role in helping farmers to counteract these problems. Planting trees along contours can be an effective way of reducing water erosion. It is substantially cheaper than conventional terracing techniques and requires much less labour. Tree planting can also be useful in helping to check gulley erosion. Windbreaks can play a similar role in helping to combat wind erosion and wind damage. Trees do not have to be arranged in discrete blocks to be effective; trees dotted around fields are often equally good provided there are enough of them. Windbreaks and shelterbelts tend to be preferred, however, especially when there is some degree of mechanization.

Although there are obvious long-term benefits from erosion control, farmers often need a short-term incentive too. Production of fruit, fodder, wood and other crops can therefore be important elements in the overall system.

NEW USES OF TREES

A third broad strategy is to encourage farmers to use trees for new purposes. A variety of different approaches may be relevant, depending on local conditions.

Soil improvement

Traditional farming systems in many areas are being progressively undermined by reduced soil fertility and deteriorating soil structure. In the past, the problem of exhaustion of the soil has been dealt with by moving to a new area and leaving the land to recover. Nowadays, increased land pressure is forcing farmers into more intensive use of their land, and sometimes even to switch to permanent cropping. For farmers with access to chemical fertilizers this may not pose too much of a problem. For the majority who have no such access, it represents a serious danger.

Nitrogen-fixing trees may be able to provide at least a partial solution. Planted along field boundaries, within fields, and in various other arrangements, they can play an important role in fertilizing and replenishing the soil.

Being perennials, trees have a number of advantages over nitrogen-fixing annuals (such as azolla, clover and alfalfa). For example:

- Perennials are often active throughout the year, absorbing water and nutrients beyond the foodcrop-growing season, thereby increasing the total productivity.
- The root system of perennials absorbs water and nutrients from deeper soil horizons, thus reducing competition with crops.
- Trees also offer protection against water and wind erosion.
- Trees can provide other benefits such as fruit, fuel, fodder and construction wood.

In one set of trials in Tanzania, nitrogen fixation in a *Leucaena* woodlot was estimated to be equivalent to 110 kg of nitrogenous fertilizer per hectare per year. Trees can also add to levels of phosphorus, potassium and other nutrients in the topsoil through their action as "nutrient pumps", absorbing nutrients from lower levels in the soil and recycling them through leaf decay in the surface layers.

A wide range of leguminous trees can be used for soil improvement, both indigenous species and exotics. To date, however, there has been very little research into the best techniques to use. Only *Leucaena* has been studied in any detail.

Information on trees suitable for arid zones is particularly lacking. In drier areas, the potential for intercropping is limited to some extent by competition between the tree and the crop for water. With some species, however, this is less of a problem. *Acacia albida*, for example, sheds its leaves during the rainy season – thus reducing its rate of transpiration and water uptake. Other trees with apparent potential in a dry-land context include *Acacia seyal*, *Vitellaria paradoxum*, *Balanites aegyptica*, *Parkia biglobosa* and *Adansonia digitata* (the baobab).

Trees for fodder production

With expanding animal populations, and reduced grazing areas, fodder provision is an increasing problem in many SADCC countries. For farmers with private land, planting trees deliberately for fodder production is one way out.

Fodder trees can play a particularly useful role in zero-grazing systems, providing a valuable protein supplement to animal diets. This is a practice that is being successfully encouraged in one agroforestry project in Lushoto District in northern Tanzania.

The best method of harvesting tree fodder will depend on the local conditions. Allowing animals to browse trees directly is the simplest method, but this can stunt the growth of the trees if not controlled carefully. Higher yields can usually be obtained by harvesting fodder by hand. This is much more labour-intensive, however, and will rarely be possible in areas of labour shortages.

Using trees for pest management

A number of tree species produce chemical compounds that can play a useful function in pest management. One common technique is to use wood ash as a protection against termites in tree nurseries. A number of other examples are listed in Table 1.

Table A1 Tree Species with Pesticidal Properties

Tree Species	Controls
Gliricidia sepium	Southern army worm, cabbage looper, corn earworm.
Anacardium occidentale	Grain weevil.
Lantana camara	Black bean aphid, rice weevil.
Melia azedirach	Corn earworm, migratory locust, brown planthopper, rice gall midge and others.
Azadirachta indica	Corn earworm, cotton stainer, migratory locust, brown planthopper, fall armyworm, rice weevil, tobacco caterpillar, etc.

The neem tree (*Azadirachta indica*) is one of the most remarkable among the list of more than 1600 species that have been identified with pesticidal properties. Traditionally, neem leaves are placed in granaries to reduce grain storage losses. A number of neem-based pesticides have also been introduced in the industrial countries. It goes without saying that neem and other natural pesticides will never fully replace chemical pesticides. However, integrated pest management incorporating both natural and chemical pesticides is receiving increasing attention and appears to have considerable potential.

NEW MANAGEMENT TECHNIQUES

A final set of approaches involves introducing farmers to new tree-management techniques. None of these techniques, in themselves, is particularly revolutionary. They are all part of traditional practices in some areas. However, for many farmers they do represent a departure from their familiar methods. In some cases they can help considerably in assisting them to raise better seedlings, more economically, and get more from the trees they plant. Appendix III discusses the various options for seedling production and the advantages and disadvantages of each.

Alternative propagation techniques

Planting out seedlings is not the only way of establishing trees. There are a number of other approaches that are often simpler and cheaper, and which may work out better for some species.

Direct Seeding

This method is commonly used in agriculture, but much less so in forestry. For farmers, and for extension agencies (see Appendix IV: Seeds versus Seedlings), it has a variety of advantages.

Depending on the species, and the local climate and soils, it can be just as effective as raising and planting out seedlings, and is generally a lot cheaper. Direct seeding of fruit trees is already commonly used by many farmers, especially women.

Adequate supplies of seed are an obvious prerequisite. When new species are being introduced, it may be necessary to establish seed orchards as an interim measure. One advantage of this method is that in places where women are forbidden by tradition from planting seedlings, it may be possible for them to establish trees by direct seeding instead.

Cuttings

Cuttings are rarely used in plantation forestry, but, in some areas, are a popular traditional method of tree propagation. Hedges, in particular, can often be successfully established through cuttings. *Lantana camara* and *Euphorbia tirucalli* are frequently planted in this way. Good techniques are required to achieve consistent results – care is needed in making the cuttings, and timing of planting is often crucial.

Wildings

"Wildings" or wild seedlings, produced by natural seeding, can be dug up by farmers and used for planting on their farms. It costs nothing, but must be done carefully and at the right time of year if good survival rates are to be achieved. Digging around seedlings a few weeks before transplanting, to prune the roots and encourage dense root growth, is one useful tip. Pruning the above-ground portion is also helpful, as it cuts down water demand and prevents seedlings from drying out during the establishment phase.

Coppicing and pollarding

The selective pruning of trees can be a useful technique for increasing yields of both wood and fodder. Coppicing involves cutting the tree close to ground level and letting it regrow from the stump. Because the root system is left alive, subsequent growth can be as much as 15 per cent faster than during the first rotation.

Pollarding involves cutting off the crown of the tree, leaving it to send out new branches from the remaining stem. A particular advantage is that the new growth is out of the reach of animals. Since the main stem is left intact, it will continue to grow and to serve as a permanent live hedge.

In both cases, careful cutting is needed to ensure healthy regrowth. Selective trimming of the resulting shoots can also help to improve the quality of the stems produced. In this way, a farmer can choose whether to keep just a few strong stems for pole production, or a large number of small stems for firewood or fodder.

Appendix II
Eucalyptus: Friend or Foe?

Eucalyptus, a genus of some 700 different tree species, has been introduced in Africa more widely than any other group. Whether this has been a good idea or not has been the subject of considerable controversy, and much heated debate. In many cases, discussion of the ecological consequences of planting eucalyptus have been mixed up with social and political arguments. Here, the major ecological issues surrounding the debate are summarized, under five main headings:

Water Demand:
: The biggest reason for planting eucalyptus in most cases is because they grow faster than many other species under the same conditions. Rapid growth is always associated with high consumption of water. The question is which is most important under the specific local circumstances – wood or water?

Climate:
: Eucalyptus plantation may have some effect on local climate, but conclusive evidence is lacking. The same is true for other tree plantations and for natural forests.

Water Run-off:
: Water run-off under eucalyptus is greater than from grassland or low shrub vegetation. Grass cover tends to be sparse, especially in dry areas and where trees are planted at close spacings. But this is also true for a lot of other species.

Soil Fertility:
: On a site that was previously treeless, eucalyptus can be expected to improve soil fertility by increasing humus levels. Its impact can be compared favourably with most coniferous species, but is less pronounced than with other broad-leaved species, especially leguminous trees.

Species Diversity:
: Eucalyptus plantations do not support a wide variety of animal or bird life. Indigenous forests are far better in this respect. If eucalyptus plantations are established, special care should be taken to conserve patches and

corridors of indigenous vegetations, especially along watercourses, as this will help maintain species diversity.

The overall conclusion that is now being increasingly accepted is that many of the negative ecological effects attributed to eucalyptus should not be blamed on the tree itself, but on the management system used to grow it – which often involves high density, short rotation, single species plantations. If this were changed, the effects of eucalyptus would be much more benign.

Appendix III
Beware The "Miracle Tree"

The danger of relying too heavily on one "miracle tree" looms for any exotic species which shows exceptionally good growth under certain conditions. The species *Leucaena leucocephala* (or "ipil-ipil", as it is sometimes known) demonstrates this point well.

Leucaena is a leguminous tree which originated in Latin America, but which has been widely introduced in southeast Asia with highly favourable reports. Under optimal conditions it grows very fast, improves the fertility of the soil and produces a whole range of useful products – including fodder, edible pods, and high quality wood.

In recent years, *leucaena* has been introduced to many African countries. Expectations have been extremely high, but in many cases it has failed to perform well. The problem is that it has often been planted without regard to local site conditions, something that no tree species can successfully survive.

For *Leucaena leucocephala*, the following general rules need to be followed if reasonable growth rates are to be achieved:

- Mountain areas should be avoided (altitudes below 1,000 m are preferred).
- Rainfall of between 600 mm and 2,500 mm per year is needed, preferably above 800 mm/year.
- Waterlogging at any part of the year should be avoided.
- Acid soils should be avoided, and lateritic rainforest soils (pH 4·0–4·5) are definitely unsuitable. Savanna soils (e.g. Alfisols) are better.
- *Leucaena* can successfully be intercropped with maize, and probably with certain other crops too. New combinations, however, need to be carefully tested before major programmes are mounted.

For other species, the requirements are different. Choosing the right tree for the site is always an essential part of designing a tree-growing programme.

Appendix IV
Options For Seedling Production

A. GOVERNMENT NURSERIES

Government nurseries, usually run by the forestry department, are the standard way of raising seedlings in most programmes. They tend to be large (100,000 seedlings or more per year) and well equipped, with pumped water, fencing, a pricking out shed, pesticides, etc. Seedlings are generally provided free, or at a highly subsidized price.

Advantages:
- Good quality-control possible.
- Easy to supervise.
- Justifies investment in pumped water on dry sites.

Disadvantages:
- Very expensive due to high cost of materials, labour, supervision.
- Seedling distribution problems limit the area they can serve.

B. GROUP NURSERIES

Group nurseries can be run by schools, women's groups, youth clubs, church groups, and many other types of organization. Scale of production can be anything from a few hundred seedlings to several thousand per year. Seedlings may be distributed among participants, or sold to generate cash.

Advantages:
- Cheaper than large centralized nurseries.
- Can act as a focus for community efforts.
- Useful educational role.

Disadvantages:
- Needs well-motivated group.
- May be difficult to maintain continuity, for example in school holidays.
- Sharing of costs and benefits may cause problems.

C. FARM NURSERIES

Tree nurseries on farms are a traditional feature in some areas. They tend to be very small, producing perhaps 100 seedlings a year, either for planting on the farm, or in some cases for sale in local markets.

Advantages:
- Extremely low cost alternative.
- Distribution problem solved.
- Maximum control in the hands of the farmer.
- Less need for outside help.

Disadvantages:
- Seedling quality depends on skill of farmer, and may be poor.
- Water supply a common problem.
- Introduction of new species and other innovations a slow process.

Appendix V
Seeds Versus Seedlings

Up to now, rural forestry projects have relied almost entirely on providing planting stock in the form of tree seedlings. This has a number of advantages. But it also has disadvantages, and in recent years there has been an increasing shift towards the idea of providing seeds direct to farmers for their own nurseries – or to plant out directly. Here, the merits of the two methods are compared:

Advantages of Seedlings
- Better able to stand competition from weeds.
- Quicker in reaching a size where they are safe from drought, fire and browsing.
- Can be raised by trained staff under controlled conditions.
- Special treatment can be provided for difficult species.

Advantages of Seeds
- Far cheaper to produce and distribute, so survival rates less crucial.
- Much easier to store and transport.
- Timing of distribution less crucial.
- Greater transfer of responsibility to the local level, and less need for long term outside support.

Disadvantages of Seedlings
- Expensive to raise, especially in government-run nurseries.
- Bulky and expensive to transport.
- Easy to damage in transit.
- Timing of distribution is crucial if seedlings are to be available at the right moment for planting.
- Good nursery techniques are essential otherwise survival rates are low.

Disadvantages of Seeds
- Unsuitable in cases where seeds are expensive or hard to obtain.
- Unsuitable for some species that require special pre-treatment to ensure good germination.
- Lower survival rate with direct seeding, especially under dry or difficult site conditions.
- Local people need experience or special training in nursery techniques if they are to raise seedlings themselves.

Select Bibliography

There are a number of journals available which contain much useful information on the themes developed in this volume.

These include: *Agroforestry Systems; SADCC Energy; ODI Social Forestry Network Papers; ICRAF Working Papers; Ambio.*

AEG and Forestry Department, *Know Your Trees (Common Trees in Zambia)*, Ndola: AEG/Forestry Department, 1979.

B. Agarwal, *Cold Hearths and Barren Slopes. The Woodfuel Crisis in the Third World*, London: Zed Books, 1986.

F.S. Alidi, "Afforestation of marginal lands for commercial timber production and to meet the needs of local communities in Botswana", in *South African Forestry Journal*, September 1984.

J. Allen, *Fuelwood Policy Options for Swaziland's RDA Programme: A Summary of Five Months In-Country Research*, New York: Cornell University Ithaca, 1985.

J.C. Allen, "Wood energy and preservation of woodlands in semi-arid developing countries: The case of Dodoma region, Tanzania", in *Journal of Development Economics*, 9:1–2, 1985.

J.C. Allen, "Soil properties and fast-growing tree species in Tanzania", in *Forest Ecology and Management*, 16, 1986.

D. Anderson, "Declining tree stocks in African countries", in *World Development*, 14:7, 1986.

D. Anderson and R. Fishwick, *Fuelwood Consumption and Deforestation in African Countries*, Washington DC: IBRD, 1984.

E. Ardayfio, *The Rural Energy Crisis in Ghana: Its Implications for Women's Work and Household Survival*, Geneva: ILO, 1986.

J.E.M. Arnold, "Replenishing the world's forests: Community forestry and meeting fuelwood needs", in *Commonwealth Forestry Review*, 62:3, 1983.

J.W. Arntzen, *Fuelwood Collection in Mosomane, Kgatleng*, Gaborone, Botswana: National Institute of Development Research and Documentation, 1983.

M.R. Bhagavan, "The energy sector in the SADCC countries", in *Ambio*, 14; 4–5, 1985.

P.F. Banks, "The potential for agro-forestry on forest land in the Sebungwe region", in *Zimbabwe Agricultural Journal*, 79:5, 1982.

G. Barnard and L. Kristofersen, *Agricultural Residues as Fuel in the Third World*, London: Earthscan, 1985.

C. Barnes, J. Ensminger and P. O'Keefe (eds), *Wood Energy and Households: Perspectives on Rural Kenya*, Uppsala: Beijer Institute and Scandinavian Institute of African Studies, 1984.

D. Barnes, *Understanding Fuelwood Prices in Developing Nations*, Washington DC: World Bank, Agriculture and Rural Development Department (unpublished), 1986.

E. Baum, "Opportunities and constraints of small-scale farms to adopt agroforestry methods in the W.Usambaras, Tanzania", in *Der Tropenlandwirt*, 85, 1984.

W.C. Beets, *Agroforestry in African Farming Systems*, Washington DC: Energy/Development International, 1985.

Beijer Institute, *Policy Options for Energy and Development in Zimbabwe*, Stockholm: Beijer Institute, 1985.

M. Best, *The Scarcity of Domestic Energy: A Study in Three Villages*, Cape Town: SALDRU Working Paper No. 27, 1979.

F. Bohlin and G. Larsson, *Village Tree Plantation Study Carried out for IRDP, Eastern Province, Zambia*, Uppsala: Swedish University of Agricultural Sciences, Working Paper 9, 1983.

F. Bohlin and G. Larsson, *Planning of Forestry for Rural Development (Work Report)*, Uppsala: Swedish University of Agricultural Sciences, 1984.

P.N. Bradley, N. Chavangi and B. van Gelder, "Development research and energy planning in Kenya" in *Ambio*, 14:4-5, 1985.

L. Buch (ed.), *Proceedings of Kenyan National Seminar on Agroforestry*, Kenya: International Council for Research in Agroforestry, 1980.

J. Bunster and A. Karlberg, *Algumas Caracteristicas e Propriedades da Lenha Comercializada na Cidade de Maputo*, Maputo: 1985.

J. Burley, "Choice of tree species and possibility for genetic improvement for small-holder and community forests", in *Commonwealth Forestry Review*, 59:3, 1979.

T. Byer, *Review of Household Energy Issues In Africa*, Washington DC: World Bank (unpublished), 1987.

P.G. von Carlowitz, "Some considerations regarding principles and practice of information collection on multipurpose trees", in *Agroforestry Systems*, 3:2, 1985.

J.H. Casey, "Selling agroforestry", in *Ceres*, 96, 1983.

E. Cecelski, "Energy and rural women's work: Crisis, response and policy alternatives", in *International Labour Review*, 126:1, 1987.

E. Cecelski, *The Rural Energy Crisis, Women's Work and Family Welfare; Perspectives and Approaches to Action*, Geneva: ILO, 1984.

M.M. Cernea, *Putting People First: Sociological Variables in Rural Development*, Oxford: Oxford University Press, 1985.

M. Chakanga and M. De Backer, *Wood Consumption and Resource Survey of Zambia*, Ndola: FAO/Forestry Department, 1985.

R. Chambers, *Rural Development. Putting the Last First*, London: Longmans, 1983.

R. Chambers and M. Leach, *Trees to Meet Contingencies: A Strategy for the Rural Poor?*, Brighton: Institute of Development Studies, 1986.

R. Chambers and M. Leach, *Trees to Meet Contingencies: Savings and Security for the Rural Poor*, London: ODI Social Forestry Network Paper, 5a, 1987.

R. Chambers and R. Longhurst, "Trees, seasons and the poor", in *IDS Bulletin*, 17:3, 1986.

E.N. Chidumayo and S.B.M. Chidumayo, *The Status and Impact of Woodfuel in Urban Zambia*, Lusaka: 1984.

J.M. Christensen, *Energy Survey in Zambia*, Denmark: Riso, 1985.

Courier, "Special issues on trees", *The Courier*, 95, 1986.

J.S. Crush and O. Namasasu, "Rural rehabilitation in the Basotho Labour Reserve", *Applied Geography*, 5, 1985.

C. Dancette and J.F. Poulain, "Influence of Acacia albida on pedoclimatic factors and crop yields", in *African Soils*, 14:1-2, 1969.

I. Dankelman and J. Davidson, *Women and Environment in the Third World*, London: Earthscan publications, 1988.

Departamento de Novas e Renovaveis Fontes de Energia, *Lenha e Carvao Vegetal na Republica Popular de Angola*, Luanda: Ministerio da Energia e Petroleos, 1987.

Departamento de Novas e Renovaveis Fontes de Energia, *Balance de Actividades*, Luanda: Ministerio da Energia e Petroleos, 1988.

Direccao Nacional de Florestas e Fauna Bravia, *A Questao da Energia Lenhosa em Mocambique. Politicas e Abordagens*, Maputo: Ministerio da Agricultura, Documento preparado para o Seminario Angola – Mocambique Sobre lenha e carvao vegetal, 1988.

R.F. Du Toit, B.M. Campbell, R.A. Haney and D. Dore, *Wood Usage and Tree Planting in Zimbabwe's Communal Lands*, Harare: Forestry Commission of Zimbabwe and World Bank, 1984.

E. Eckholm, *The Other Energy Crisis: Firewood*, Washington DC: Worldwatch Institute, 1975.

J.A. Eklofaud and H. Petterson, *A Study on Energy Use and Afforestation in Tabora*, Dar es Salaam: 1984.

Energy Studies Unit (ESU), *Malawi Rural Energy Survey*, Lilongwe: ESU, 1982.

Energy Studies Unit (ESU), *Malawi Urban Energy Survey*, Lilongwe; ESU, 1984.

Energy Studies Unit (ESU), *Flue-Cured Tobacco Energy Use Efficiency Survey*, Lilongwe: ESU, 1985.

Environmental Resources Limited (ERL), *A Study of Energy Utilisation and Requirements in the Rural Sector of Botswana: Final Report*, London: ERL/ODA, 1985.

ETC, *Wood Energy Development: Policy Issues. A Study of the SADCC Region*, Leusden: SADCC Energy Development Fuelwood Study, ETC, 1987.

ETC, *Wood Energy Development: A Planning Approach*, Leusden: SADCC Energy Development Fuelwood Study, ETC, 1987.

ETC, *Wood Energy Development: Biomass Assessment*, Leusden; SADCC Energy Development Fuelwood Study, ETC, 1987.

ETC, *Wood Energy Development: the LEAP Model*, Leusden: SADCC Energy Development Fuelwood Study, ETC, 1987.

ETC, *Wood Energy Development: Bibliography of the SADCC Region*, Leusden: SADCC Energy Development Fuelwood Study, ETC, 1987.

D.B. Fanshawe, *Useful Trees of Zambia for Agriculture*, Lusaka: Division of Forestry Research, 1972.

FAO, *Institutional Aspects of Shifting Cultivation in Africa*, Rome: FAO, 1984.

FAO, *Tree Growing by Rural People*, Rome: FAO, 1985.

FAO, *Wood Based Energy and Substitution among Fuels in Africa; Model Framework and Basic Data from Eleven Country Reports*, Rome: FAO, 1987.

FAO, UNEP, *Tropical Forest Resources Assessment Project. Forest Resources of Tropical Africa: Part 1 Regional Synthesis: Part 2 Country Briefs*, 2 vols, Rome: FAO, 1981.

Fernandes et al, "The Chagga homegardens: a multi-storied agroforestry cropping system on Mt. Kilimanjaro", in *Agroforestry Systems*, 2:2, 1984.

J.H. Ferreira de Castro, *Forest Resources in Mozambique and their Rational Use*, Rome: FAO, 1978.

P.C. Fleuret and A.K. Fleuret, "Fuelwood use in a peasant community: A Tanzanian case study", in *Journal of Developing Areas*, 12, 1978.

G. Foley, *Charcoal Making in Developing Countries*, London: Earthscan, 1986.

G. Foley and G. Barnard, *Farm and Community Forestry*, London: Earthscan, 1984.

G. Foley, P. Moss and L. Timberlake, *Stoves and Trees*, London: Earthscan, 1984.

Forest Research Institute, *A Preliminary Silvicultural Classification of Malawi*, Zomba: Forest Research Institute of Malawi, 1978.

Forest Division (Tanzania), *Trees for Village Forestry*: Dar es Salaam: Ministry of Lands Natural Resources and Tourism, Swedish International Development Agency, 1984.

L. Fortmann and J. Riddell, *Trees and Tenure: An Annotated Bibliography for Agroforesters and Others*, Madison, Wisconsin: Land Tenure Centre, University of Wisconsin, 1985.

D. French, "Confronting an unsolvable problem: deforestation in Malawi", in *World Development*, 14:4, 1986.

D. French, *Agroforestry for Food, Fuel or Income: Focusing on Women's Work*, Rome: FAO, 1986.

J. Gay, *Lesotho Household Energy Survey - 1984*, Maseru: Ministry of Cooperatives and Rural Development, 1984.

J. Gay and M. Khoboko, *Village Energy Survey Report*, Maseru: Ministry of

Cooperatives and Rural Development, 1982.

J. Gill, "Improved stoves in developing countries: a critique", in *Energy Policy*, 15, 1987.

A. Greaves, "Gmelina arborea", in *Forestry Abstracts*, 42, 1981.

H.M. Gregersen, S. Draper, D. Elz (eds), *People and Trees: Social Forestry Contributions to Development*, Washington DC: World Bank, in press.

B.C. Groen and C.R. Huizenza, *Have Planners Understood the Poor People's Energy Problem? Socio-economic aspects of Energy Technologies; A Literature Review*, Enschede: Technology and Development Group, University of Twente, 1987.

M. Grut, *Guidelines for Identifying and Preparing Forestry Projects*, Washington DC: IBRD, 1986.

GTZ, *Lesotho Energy Study*, GTZ, 1984.

G.M. Guess, "Technical and financial policy options for development forestry", in *Natural Resources Journal*, 21, 1981.

D.A. Gwaitta-Magumba, "Afforestation of marginal lands for commercial timber production and to meet the needs of the rural community in Swaziland", in *South African Journal of Forestry*, September 1984.

I. Hague and S. Jutzi, "Nitrogen fixation by forage legumes in Sub-Saharan Africa: potential and limitations", in *ILCA Bulletin*, 20, 1984.

D.O. Hall and P. Moss, "Biomass for energy in developing countries", in *Geojournal*, 7:1, 1983.

D.O. Hall, "Biomass: Fuel versus food, a world problem?", in M.S. Margaris (ed.), *Economics of Ecosystem Management*, Netherlands: W. Junk Publishers, 1984.

L.S. Hamilton and P.N. King, "Watersheds and rural development planning", in *Environmentalist*, 4:7, 1984.

D. Hancock and G. Hancock, *Cooking Patterns and Domestic Fuel Use in Masvingo Province*, GTZ, 1985.

D.D. Hardcastle, "A preliminary silvicultural classification of Malawi", in *Malawi Forestry Record*, 57, 1977.

P. Harou, W.A. Patterson and J. Falconi, "The role of forestry in dry Africa", in *Journal of Forestry*, 83:3, 1985.

P. Harrison, *The Greening of Africa*, London: Paladin, Grafton Books, 1987.

O. Hofstad, *Preliminary Evaluation of the Taunya System for Combined Wood and Food Production in North-Eastern Tanzania*, Dar es Salaam: Division of Forestry, University of Dar es Salaam, 1978.

P. Hogberg and M. Kvarnstrom, "Nitrogen fixation by the woody legume Leucaena in Tanzania", in *Plant and Soil*, 66, 1982.

M. Horowitz and K. Badi, *Sudan, Introduction of Forestry in Grazing Systems*, Rome: FAO, 1981.

F. Horsten, *Madeira – Uma Analise da Situacao Actual*, Luanda: Ministry of Agriculture, 1983.

R.H. Hosier, Y. Katerere, D.J. Munasirei, J.C. Nkomo. B.J. Ram, and P.B. Robinson, *Zimbabwe: Energy Planning for National Development*, Uppsala:

Scandinavian Institute of African Studies, 1986.

M.W. Hoskins, *Women in Forestry for Local Community Development*, Washington DC: Office of Women in Development, AID, 1979.

M. Howes, *Rural Energy Surveys in the Third World: A Critical Review of Issues and Methods*, Canada: International Development Research Centre, 1985.

P.A. Huxley, 'Experimental agroforestry-progress through perception and collaboration?", in *Agroforestry Systems*, 3:2, 1985.

ICRAF, *A Selected Bibliography of Agroforestry*, Nairobi: International Council for Research in Agroforestry, 1982.

ICRAF, *Agroforestry Research and Development: ICRAF at Work*, Nairobi: International Council for Research in Agroforestry, 1987.

ICRAF, *D and D User's Manual: An Introduction to Agroforestry Diagnosis and Design*, Nairobi: International Council for Research in Agroforestry, 1987.

Institute of Adult Education and Ministry of National Resources and Tourism, *Final Report of "Forests are Wealth" Campaign*, Tanzania: IAE and MNRT, 1982.

International Tree Project Clearing House (ITPC), *Directory of Forestry NGOs in Africa*, New York: UN NGO Liaison Service, ITPC, 1987.

M.C.R. Jacinto, *Producao e Consumo de Carvao e Lenha em Maputo*, Maputo: Universidade Eduardo Mondlane, 1978.

S.A.A. Jahn, H.A. Musnad and H. Burgstaller, "The tree that purifies water: cultivating multipurpose Moringaceae in the Sudan", in *Unasylva*, 152:38, 1986.

N.E. Jelenic and J.A. Van Vegten, *A Pain in the Neck: The Firewood Situation in South-Western Kgatleng, Botswana*, Gaborone: University of Botswana, 1981.

B.K. Kaale, "Planning guidelines for village afforestation in Tanzania (summary)", in *Forestry Abstracts*, 44:10, 1981.

B.K. Kaale, *Trees for Village Forestry*, Dar es Salaam: Forest Division, 1984.

B.K. Kaale, *Guidelines for Compiling Regional and District Afforestation Plans in Tanzania*, Dar es Salaam: Ministry LNRT, 1985.

D.M. Kamweti, *Fuelwood in East Africa: Present Situation and Prospects*, Rome: FAO, 1984.

B.T. Kang, G.F. Wilson, T.L. Lawson, *Alley Cropping: A Stable Alternative to Shifting Cultivation*, Ibadan: International Institute of Tropical Agriculture, 1985.

Y. Katerere, "The fuelwood crisis: some possible solutions to Zimbabwe's burning issues", in *Zimbabwe Science News*, 19, 1985.

D.L. Kgathi, *Aspects of Firewood Trade between Rural Kweneng and Urban Gaborone: A Socioeconomic Perspective*, Gaborone: National Institute of Development Research and Documentation, 1984.

A. Kir, *Retrospectiva do Sector Florestal e Linhas Gerais de Desenvolvimento*, Maputo: Ministry of Agriculture/FAO, 1984.

H. Kjekshus, *Ecology Control and Economic Development in East African History*, London: Heinemann, 1977.

G.S. Kowero and A.B. Temu, "Some observations on implementing village forestry programmes in Tanzania", in *International Tree Crops Journal*, 3, 1985.

G. Kuchelmeister, *State of Knowledge Report on Tropical and Subtropical Hedgerows*, Eschborn: German Agency for Technical Cooperation, 1987.

M.S. Kumar (ed.), *Energy Pricing Policies in Developing Countries*, Geneva: ILO, 1987.

H.N. Le Houerou, *Browse in Africa: The Current of Knowledge*, Addis Ababa: International Livestock Centre for Africa, 1980.

IITA, *Alley Cropping – A Stable Alternative to Shifting Cultivation*, Ibadan: IITA, n.d.

T. Laya-Sensenig, *Let's Solve our Firewood Problem: Stoves and Trees*, Morogoro: University of Dar es Salaam, 1984.

G. Leach, "Energy and the urban poor", in *IDS Bulletin*, 18:1, 1987.

G. Leach and M. Gowen, *Household Energy Handbook*, Washington DC: World Bank, 1987.

G. Leach and R. Mearns, *Bioenergy Issues and Options*, London: Report to the Royal Norwegian Ministry of Development Cooperation, IIED, 1988.

L. Lekalake, *Women and Development in Botswana: An Annotated Bibliography*, Gaborone: Ministry of Home Affairs, 1985.

T. Mabbett, "Lifting the smokescreen on wood use by tobacco growers", in *Agriculture Administration*, 39:3, 1987.

J. Mabonga-Kwiraka, *A Report on Charcoal Production in Maputo*, Rome: FAO, 1978.

J.A. Maghembe and J.F. Redhead, *Agroforestry: Preliminary Results of Intercropping Acacia, Eucalyptus and Leucaena with Maize and Beans*, Morogoro: Proceedings of a conference on intercropping, 1982.

F. Malaisse and K. Binzangi, "Wood as a source of fuel in Upper Shaba, Zaire", in *Commonwealth Forestry Review*, 64:3, 1985.

J. Malleux, *Evaluacion de los Recursos Forestales de la Republica Popular de Mozambique*, Rome: FAO, 1981.

E. Mansur and A. Karlberg, *Levantamento do Abastecimento de Lenha e Carvao na Cidade de Maputo*, Maputo: Ministry of Agriculture, 1986.

R.B. Martin, *Communal Areas Management Programme for Indigenous Resources (CAMPFIRE)*, Harare: Department of National Parks and Wildlife Management, 1986.

A. Mascarenhas, I. Kikula and P. Nilsson, *Support to Village Afforestation in Tanzania*, Dar es Salaam: Institute of Resource Assessment, University of Dar es Salaam, 1983.

A. Mascarenhas, L.A. Odero-Ogwel, Y.F.O. Masakhalia and A.K. Biswas, "Land use policies and farming systems: Kenya, Tanzania and Mozambique", in *Land Use Policy*, 3:4, 1986.

E. Maudoux, *Planification Forestierre Angola*, Rome: FAO, 1984.

D. Mazambani, "Exploitation of trees around Harare", in *Zimbabwe Science News*, 16:11, 1982.

D. Mazambani, "Peri-urban deforestation in Harare", in *Proceedings of the Geographical Association of Zimbabwe*, 14, 1983.

J. McClintock, *Fuelwood Scarcity in Rural Africa: Possible Remedies*, Vol. 1, Reading: Centre for Agricultural Strategy, University of Reading, 1987.

R. Melamed-Gonzalez and L. Giasson, *A Directory of NGOs in the Forestry Sector*, 2nd African edition, New York: International Tree Project Clearinghouse, 1987.

A.S.M. Mgeni, "Fuelwood crisis in Tanzania is women's burden", in *Quarterly Journal of Forestry*, 73:4, 1984.

A.S.M. Mgeni, "Soil conservation in Kondoa District Tanzania", in *Land Use Policy*, 2:3, 1985.

A. Millington and J. Townsend, *Biomass Assessment. A Study of the SADCC Region*, London: Earthscan Publications 1988.

Ministry of Mineral Resources and Water Affairs (MRWA), *Botswana Energy Masterplan*, Gaborone: Ministry of MRWA, 1986.

E.M. Mnzava, *Village Afforestation: Lessons of experience in Tanzania*, Rome: FAO, 1980.

E.M. Mnzava, "Tanzanian tree planting: A voice from the villagers", in *Unasylva*, 1985.

B. Munslow and P.O'Keefe, "Energy and the regional confrontation in Southern Africa", in *Third World Quarterly*, 6:1, 1984.

B. Munslow, P. Phillips. D. Pankhurst and P. O'Keefe, "Energy and Development on the African East Coast: Somalia, Kenya, Tanzania and Mozambique", in *Ambio*, 12:6, 1983.

M.J. Mwandosya and M.L.P. Luhanga, "Energy use patterns in Tanzania", in *Ambio*, 14, 4–5, 1985.

P.K.R. Nair, "Classification of agroforestry systems", in *Agroforestry Systems*, 3:2, 1985.

P.K.R. Nair, *Agroforestry Systems in Major Ecological Zones of the Tropics and Subtropics*, Nairobi: ICRAF Working Paper No. 47, 1987.

P. Neunhauser, E. Hauser, D. Aehling, R. Droste, C. Graefen, H. Kaya, R. Schmitt, H. Stamm and K. Wagner, *Demand for Major Fruit Tree Seedlings Including Coconut by Village Farms and Farmers in the Lowland Areas of the Tanga Region*, Berlin: Centre for Advanced Training in Agricultural Development, Technical University of Berlin, 1986.

R.S.W. Nkaonja, "Rural fuelwood and poles research project in Malawi: a general account", in *South African Forestry Jorunal*, 117, 1981.

R.W.S. Nyirenda, *The Economics of Wood Energy Production for Tobacco Processing in Malawi: A case study*, Bangor: University College of North Wales, Dept. of Forestry & Wood Sciences (unpublished MSc thesis), 1983.

ODA/ERL, *A Study of Energy Utilization and Requirements in the Rural Sector of Botswana*, London: ODA/ERL, 1985.

P. O'Keefe and B. Munslow, *Energy and Development in Southern Africa. SADCC Country Studies* (2 vols), Uppsala: Beijer Institute and Scandinavian Institute of African Studies, 1984.

P. O'Keefe, P. Raskin and S. Bernow (eds), *Energy and Development in Kenya: Opportunities and Constraints*, Uppsala: Beijer Institute and Scandinavian Institute of African Studies, 1984.

A. O'Kting'Ati, *Fuelwood Consumption in an Economically Depressed Urban Centre (Tabora)*, Dar es Salaam: Forest Division, 1984.

A. O'Kting'Ati et al, "Plant species in the Kilimanjaro agroforestry system", in *Agroforestry Systems*, 2:3, 1984.

W. Ostberg, *The Kondoa Transformation: Coming to Grips with Soil Erosion in Central Tanzania*, Uppsala: Research Report 76, Scandinavian Institute of African Studies, 1986.

R. Peet, *Manufacturing Industry and Economic Development in the SADCC Countries*, Uppsala: Beijer Institute and Scandinavian Institute of African Studies, 1984.

J. Persson, "Trees, plants and a rural community in the southern Sudan", in *Unasylva*, 154, 1986.

D. Poore and C. Fries, *The Ecological Effects of Eucalyptus*, Rome: FAO, 1985.

P. Poschen, "An evaluation of the Acacia albida-based agroforestry practices in the Hararghe highlands of eastern Ethiopia", in *Agroforestry Systems*, 4:2, 1986.

G. Poulson, *Malawi: The Function of Trees in Small Farmer Production Systems*: Rome: FAO, 1981.

G. Poulson, "Using farm trees for fuelwood", in *Unasylva*, 35, 1984.

B. Pound and L. Martinez Cairo, *Leucaena – its Cultivation and Uses*, London: ODA, 1983.

P. Powell and P. Wellings, "The Lesotho Woodlot Project: progress, problems and prospects", in *Development Studies for Southern Africa*, 5, 1983.

J.B. Raintree and K. Warner, "Agroforestry pathways for the intensification of shifting cultivation", in *Agroforestry Systems*, 4:1, 1986.

J.B. Raintree, "Agroforestry pathways: Land tenure, shifting cultivation and sustainable agriculture", in *Unasylva*, 154, 1986.

P. Richards, *Indigenous Agricultural Revolution*, London: Hutchinson, 1985.

SADCC Energy Sector, *Woodfuel Seminar*, Luanda: SADCC Energy Sector, 1983.

C. Sana, *Shinyanga Forestry Project: Evaluation Report*, Shinyanga: Oxfam, 1984.

A.G. Seif El Din, *Integrated Land Use in Forest Reserves in Eastern Region: Global Diagnosis and the Involvement of the People*, Khartoum: Government of the Sudan and Government of the Netherlands, FAO, 1986.

M.W.M. Shaba, *Deforestation and Land Use in Malawi with Special Reference to the Southern Region*, Bangor: University College of North Wales (unpublished MSc thesis), Department of Forestry & Wood Science, 1984.

S.N. Siame, *Guideline on Community and Agroforestry Impacts and Integration with Plantation Forestry*, Ndola: Forestry Department, 1985.

J.C.T. Simoes (ed.), *SADCC Energy and Development to the Year 2000*, Uppsala: Beijer Institute and Scandinavian Institute of African Studies, 1984.

M.M. Skutsch, "Forestry by the people for the people: Some major problems in Tanzania's village afforestation programme", in *International Tree Crops Journal*, 3, 1986.

M.M. Skutsch, "Participation of women in social forestry programmes", in *Bos Newsletter*, 13:5 (1), 1986.

M.M. Skutsch, K.F. Wiersum and J. Waite, *Kenya Woodfuel Development Programme Mid-term Review*, Nairobi: Final report for the Directorate General for International Cooperation, Ministry of Foreign Affairs, Kenya.

M.M. Skutsch, *Why People Don't Grow Trees*, Washington DC: Resources for the future, 1983.

V. Smil and W.E. Knowland, *Energy in the Developing World: The Real Energy Crisis*, Oxford: Oxford University Press, 1980.

J.S. Spears, "Can farming and forestry co-exist in the tropics?", in *Unasylva*, 32, 1980.

W. Stberg, *The Kondoa Transformation: Coming to Terms with Soil Erosion in Central Tanzania*, Uppsala: Scandinavian Institute of African Studies, 1986.

P. Stromgaard, "Biomass growth and burning of woodland in a shifting cultivation area of South Central Africa", in *Forest and Energy Management*, 12, 1985.

Swedforest Consulting AB, *Report from the Joint Mission on Village Forestry and Industrial Plantations in Tanzania*, Danderyn: Swedforest Consulting AB, 1986.

T.P. Tabkoeta, *Forestry and Landuse Improvement in Northern Cameroon*, Bangor: University of North Wales (unpublished Msc thesis), 1986.

D.N. Tembo, C. Chandrasekharan, *Understanding Tree Use in Farming Systems; Workshop on Planning Fuelwood Project with Participation of Rural People*, Rome; FAO, 1984.

A.B. Temu, B.K. Kaale and J.A. Maghembe (eds), *Wood Based Energy For Development*, Dar es Salaam: Ministry of Natural Resources and Tourism, Swedish International Development Agency, 1984.

T. Tietema, *Firewood for Botswana: Towards a Sustained Harvest of Firewood*, Harare: Unesco Zimbabwe, 1984.

L. Timberlake, *Africa in Crisis*, London: Earthscan, 1985, new edition 1988.

UNDP/World Bank, *Malawi: Issues and Options in the Energy Sector*, Washington DC: World Bank/UNDP, 1982.

UNDP//World Bank, *Zimbabwe: Issues and Options in the Energy Sector*, Washington DC: World Bank/UNDP, 1982.

UNDP/World Bank, *Zambia: Issues and Options in the Energy Sector*, Washington DC: World Bank/UNDP, 1983.

UNDP/World Bank, *Botswana: Issues and Options in the Energy Sector*, Washington DC: World Bank/UNDP, 1984.

UNDP/World Bank, *Lesotho: Issues and Options in the Energy Sector*, Washington DC: World Bank/UNDP, 1984.

UNDP/World Bank, *Tanzania: Issues and Options in the Energy Sector*, Washington DC: World Bank/UNDP, 1984.

UNDP/World Bank, *Zimbabwe Energy Assessment Status Report*, Washington DC: World Bank/UNDP, 1984.

UNDP/World Bank, *Mozambique: Issues and Options in the Energy Sector*, Washington DC: World Bank/UNDP, 1987.

UNDP/World Bank, *Tanzania Urban Woodfuels Supply Study*, Washington DC: World Bank, 1987.

Universidade Eduardo Mondlane (UEM)/ILO, *Trabalho da Mulher Rural, Utilizacao do Combustivel Domestico e Nutricao: Garcia, Andrade, Antinao e Loforte*, Maputo: UEM/ILO, 1984.

B. van Gelder, *Agroforestry Components in Existing Netherlands-Assisted Projects in Kenya*, Nairobi: ETC Foundation, 1988.

J.C. Westoby, "Foresters and politics", in *Commonwealth Forestry Review*, 64:2, 1985.

Whitsun Foundation, *Rural Afforestation Study*, Harare: Whitsun Foundation, 1981.

G.E. Wickens, J.R. Goodin and D.V. Field (eds), *Proceedings of the Kew International Conference in Economic Plants for Arid Lands*, London: George Allen & Unwin, 1985.

K.F. Wiersum (ed.), *Proceedings of the International Symposium on Strategies and Designs for Reforestation, Afforestation and Tree Planting*, Wageningen: Agricultural University, 1984.

K.F. Wiersum, *Forestry Aspects of Stabilizing Shifting Cultivation in Africa*, Wageningen: Agricultural University, 1985.

K.F. Wiersum, "Trees in agricultural and livestock development", in *Netherlands Journal of Agricultural Science*, 33, 1985.

K.B. Wilson, *Research on Trees in Masvihwa and Surrounding Areas*, Harare: ENDA-Zimbabwe report, 1987.

G.F. Wilson, B.T. Kang and K. Mulongoy, "Alley cropping: Trees as sources of green-manure and mulch in the tropics", in *Biological Agriculture and Horticulture*, 3, 1986.

B. Wisner, "Rural energy and poverty in Kenya and Lesotho: All roads lead to ruin", in *IDS Bulletin*, 18:1, 1987.

D.J. Wolfson and D. Ghai, *The Rural Energy Crisis, Women's Work and Basic Needs*, Geneva: ILO, 1986.

World Bank, *Zambia Energy Assessment Status Report*, Washington DC: World Bank, 1985.

World Commission on Environment and Development, *Our Common Future*, Oxford: Oxford University Press, 1987.

Index

Printed and bound by CPI Group (UK) Ltd, Croydon, CR0 4YY

23/10/2024

01777674-0003